THE ENLIGHTENMENT

Studies in European History

Series Editors: John Breuilly
                   Julian Jackson
                   Peter Wilson

Studies in European History
Series Standing Order ISBN 978–0333–79365–7
(outside North America only)

You can receive future titles in this series as they are published by placing a standing order. Please contact your bookseller or, in case of difficulty, write to us at the address below with your name and address, the title of the series and the ISBN quoted above.

Customer Services Department, Macmillan Distribution Ltd
Houndmills, Basingstoke, Hampshire RG21 6XS, England

The Enlightenment

Second Edition

Roy Porter

palgrave

First published 2001 by
PALGRAVE
Houndmills, Basingstoke, Hampshire RG21 6XS and
175 Fifth Avenue, New York, N.Y. 10010
Companies and representatives throughout the world

PALGRAVE is the new global academic imprint of
St. Martin's Press LLC Scholarly and Reference Division and
Palgrave Publishers Ltd (formerly Macmillan Press Ltd).

ISBN 0–333–94505–0

This book is printed on paper suitable for recycling and
made from fully managed and sustained forest sources.

A catalogue record for this book is available
from the British Library.

Library of Congress Cataloging-in-Publication Data
Porter, Roy, 1946–
    The Enlightenment / Roy Porter.—2nd ed.
    p. cm.—(Studies in European history)
Includes bibliographical references and index.
ISBN 0–333–94505–0
    1. Enlightenment—Europe.   2. Europe—Intellectual life—
18th century.   I. Title   II. Studies in European history
(New York, N.Y.)

B802. P62 2000
940.2′53—dc21   00–041507

Contents

Editor's Preface

The main purpose of this new series of studies is to make available to teacher and student alike developments in a field of history that has become increasingly specialized with the sheer volume of new research and literature now produced. These studies are designed to present the 'state of the debate' on important themes and episodes in European history since the sixteenth century, presented in a clear and critical way by someone who is closely concerned himself with the debate in question.

The studies are not intended to be read as extended bibliographical essays, though each will contain a detailed guide to further reading which will lead students and the general reader quickly to key publications. Each book carries its own interpretation and conclusions, while locating the discussion firmly in the centre of the current issues as historians see them. It is intended that the series will introduce students to historical approaches which are in some cases very new and which, in the normal course of things, would take many years to filter down into the textbooks and school histories. I hope it will demonstrate some of the excitement historians, like scientists, feel as they work away in the vanguard of their subject.

The format of the series conforms closely with that of the companion volumes of studies in economic and social history, a series that has already established a major reputation since its inception in 1968. Both series have an important contribution to make in publicising what it is that historians are doing and in making history more open and accessible. It is vital for history to communicate if it is to survive.

R. J. OVERY

A Note on References

References are cited throughout in square brackets according to the numbering in the Suggested Reading, with page reference where necessary indicated by a colon after the bibliography number.

Introduction and Acknowledgements

In the late 1960s there appeared Peter Gay's superb two-volume survey of eighteenth-century thought, *The Enlightenment: An Interpretation* [61]. In just under a thousand pages of text, Gay, professor of history first at Columbia University and then at Yale, offered a comprehensive investigation of the period, its problems, and its protagonists, written in a lucid and lively style. A further 250 pages of 'Bibliographical Essay' listed and evaluated the massive scholarship already available upon the subject.

Gay wrote at a moment when research into the Enlightenment was beginning to mushroom. In the three decades since his work was published, so many new books and articles have appeared, striking out in a multitude of new directions, and often challenging old truths, that it is doubtful if a comparable enterprise to Gay's could today be contained in fewer than four or five volumes, with a bibliography which would fill another. And who would write it?

Given the vast increase of scholarly attention lavished on the Enlightenment, it may seem the height of folly to hope to say anything worthwhile in the span of a hundred pages. But it is important to try. For very few school and college students have easy access to the mountains of new monographs and an ever-increasing number of specialist academic journals covering such diverse fields as literary studies, cultural history, social history, women's studies, the history of ideas, religion, science, and so forth. Few libraries even possess a complete run of the fundamentally important publications of the Voltaire Foundation, whose titles now number around four hundred.

In an introductory pamphlet such as this, there is no room to assess more than a fraction of all this research, or even to list it in the 'Reading Suggestions' at the end of the book. Rather, my approach has been to take the main interpretations and problems respecting the Enlightenment as set out in the leading, readily available accounts produced during the last generation – the works of Gay [59; 60; 61], Hampson [72], Crocker [44; 45], May [103], Jacob [84; 85; 86], Outram [109], Pocock [116; 117; 118] and so forth – and to gauge how valid these still remain, or how far, on the contrary, they now require modification. In some fields of study, fresh research has brought to light crucial new information. In others, our perception of what the important questions in the history of thought, ideas, and culture really are, has radically changed. This is something which advocates of Enlightenment would have appreciated: the 'Preliminary Discourse' to Diderot and d'Alembert's celebrated *Encyclopédie* spoke of the need to create new forms of knowledge to meet the needs of a new world [92; 125].

My aims have thus been expository, critical and historiographical; yet I have tried to avoid writing what would essentially be a 'review essay', a mere commentary upon recent scholarship. Instead, I have tried to produce a work which is self-standing and self-contained. Readers will note that rather little is said in the following chapters about the *ancien régime* as such, about Enlightened Absolutism, or about the origins of the French Revolution. This is not because I think they are not germane to the subject. Far from it. But these topics are already fully covered in excellent books appearing in this series, respectively by William Doyle [52] and Tim Blanning [27].

I am most grateful to the editors in this series, Richard Overy and John Breuilly, for their support, encouragement and criticisms in transforming a rough draft into a finished work. Mark Goldie, Margaret Kinnell, Sylvana Tomaselli, Jane Walsh, Andrew Wear, and Bob Wolfson also commented upon earlier drafts. Their kindness has saved me from many blunders: my warmest thanks to them. Vanessa Graham at the publishers was patient and supportive in the production of the original text ten years ago, and Terka Bagley has been equally so for this second edition, in which I have taken the opportunity to correct errors, discuss recent trends in interpretation and add some fifty items to the Bibliography. I would also like to thank the Governors of the Wellcome

Trust for providing, during the last twenty years, an excellent working environment at the Wellcome Institute for the History of Medicine. I dedicate this book to the memory of a wonderful institution.

A Note on Usage: Enlightened writers commonly used gendered language – e.g., 'man of letters'. Such usage largely reflected the realities of the age, a time when the vast majority of writers (for example) were male. For this reason I have chosen to avoid rigidly following modern politically correct usages.

1 What was the Enlightenment?

Just over two hundred years ago, the German philosopher Immanuel Kant wrote an essay entitled '*Was ist Aufklärung?*' ('What is Enlightenment?'). For Kant, Enlightenment was mankind's final coming of age, the emancipation of the human consciousness from an immature state of ignorance and error. He believed this process of mental liberation was actively at work in his own lifetime. The advancement of knowledge – understanding of Nature, but human self-knowledge no less – would propel this great leap forward. '*Sapere aude*' ('dare to know') was Kant's watchword, taken from the Roman poet Horace.

But only the most unquestioning historian today would pronounce, as confidently as Kant, that what we now know as the Enlightenment of the eighteenth century, that body of 'progressive' and 'liberal' ideas and opinions advanced by the leading intellectuals and propagandists of the day, unambiguously amounted to a decisive stage in human improvement. Historians are rightly sceptical about taking the spokesmen of the past upon their own terms. In any case, 'saints and sinners' histories, which paint pictures of forward-looking 'heroes' slaying reactionary tyrants and bigots to create a better future, nowadays themselves appear partisan and prejudiced. It would be folly to hope to find in the Enlightenment a perfect programme for human progress. It should rather be seen as posing a series of problems for historians to explore.

For long, the movement suffered a bad press, especially in Britain. The 'Age of Reason' – the portmanteau term traditionally given to eighteenth-century views – was dismissed by the Victorians as a time of shallow and mechanical thinkers, overweeningly

1

confident in the powers of abstract reason. Reason alone (Enlighteners were believed to believe) would afford a total knowledge of man, society, Nature and the cosmos; would enable them to mount a critique of the political and religious *status quo*; and, above all, would provide the foundations for a utopian future. Far more, however, existed in the world (so the Romantics later argued) than was day-dreamt about in the armchair philosophies of the Enlightenment: not least, the imagination, feeling, the organic power of tradition and history, and the mysteries of the soul. Sometimes silly, often seductive, but always shallow, Enlightenment teachings had proved appallingly dangerous. Its much-vaunted humanitarianism had led (so many Victorians accused) to the crimes against humanity committed in the French Revolution and thereafter. Unsympathetic critics, nowadays postmodernist as well as conservative, still make similar insinuations [126; 146; 160].

The 'Age of Reason' found few friends in the nineteenth century. Romantics judged it soulless, conservatives thought it too radical, while radicals in turn were distressed to find its leaders, notably Voltaire, were at bottom worldly elitists, salon talkers rather than revolutionary activists. Only in the twentieth century, when the true complexities of the relations between ideology and action have forced themselves upon us, have the subtle ironies of the Enlightenment come to be appreciated.

For one thing, all historians now agree that the very labelling of the eighteenth century as an 'age of reason' is deeply misleading [61]. Many of the century's leading intellectuals themselves dismissed the rationalist, system-building philosophers of the seventeenth century, notably Descartes (with his notion of 'clear and distinct ideas' self-evident to reason) and Leibniz. They repudiated them as fiercely as they rejected what they considered the verbal sophistries of rationalist, scholastic theology, developed first by St Thomas Aquinas in the Middle Ages (Thomism), and further elaborated in the Counter-Reformation. In the light of the triumph of Newtonian science, the men* of the Enlightenment argued that experience and experiment, not *a priori* reason, were the keys to true knowledge [135]. Man himself was no less a feeling than a thinking animal. No doubt, as Goya observed, the 'sleep of

---

* Practically all the key Enlightenment thinkers were, indeed, male. For the role of women in the Enlightenment see below, Chapter 7.

reason produces monsters'. But, divorced from experience and sensitivity, reason equally led to error and absurdity, as Voltaire delightfully demonstrated in his philosophical novel *Candide*, in which the stooge, Dr Pangloss, is so blinded by his Leibnizian metaphysical conviction that 'all is for the best in the best of all possible worlds', as to become utterly indifferent to the cruelty and suffering going on under his best of all possible noses [61: vol. 1, 197; 158].

As Gay has emphasized, the exponents of Enlightenment were neither rationalists, believing that reason was all, nor irrationalists, surrendering their judgement before feeling, faith, intuition and authority [61: vol. 1, 127f.]. They criticized all such simple-minded extremes, because they were, above all, *critics*, aiming to put human intelligence to use as an engine for understanding human nature, for analysing man as a sociable being, and the natural environment in which he lived. Upon such understanding would the foundations for a better world be laid.

They called themselves 'philosophers', and this term (in the French form, *philosophes*) will serve – for there is no exact English equivalent – as a convenient group name for them below. (Occasionally the German form, *Aufklärer* (Enlighteners), will be used.) But we must not think of them as akin to the stereotypical philosophy professor of today, agonizing over the nuances of words in his academic ivory-tower. Rather they were men of the world: journalists, propagandists, activists, seeking not just to understand the world but to change it. Diderot and d'Alembert's *Encyclopédie* thus defined the *philosophe* as one who 'trampling on prejudice, tradition, universal consent, authority, in a word, all that enslaves most minds, dares to think for himself'. Voltaire was to the fore in campaigning against legal injustice in a succession of *causes célèbres* in the 1760s; for a brief spell, it fell to the *philosophe*-economist Turgot to take charge of the French finances; the leading American intellectual, Benjamin Franklin of Philadelphia, put the science of electricity on the map, invented bifocal spectacles as well as lightning conductors, and also played a crucial part in setting up the new American Republic [17; 103].

A more rounded knowledge of such intellectuals as Diderot and Condorcet has dispelled the old caricature of the *philosophes* as dogmatic system-builders, infatuated with pet economic nostrums and 'vain utopias seated in the brain' [19; 125; 164]. Above all, we

3

should be careful not to give oversimplified accounts of their ideas. They often popularized – to get through to the people. They often sloganized (they needed to, in order to be heard). But there was much subtlety behind the slogans. From around 1760, Voltaire went onto the offensive against the evils of religion with what became a notorious catchphrase: '*Écrasez l'infâme*' (destroy the infamous one). Yet it would be simplistic to jump to the conclusion that he had declared total war on all religion whatsoever (see below, Chapter 4). Experience of twentieth-century police states should have taught us why the *philosophes* had to speak in foreign tongues under different circumstances: now they had to be blunt, now they had to hold forth in riddles or fables, in order to circumvent the all-present censor [48]. Straight-talking was not always possible or effective.

Once ingrained myths and prejudices are thus cleared away, we can begin to reassess the nature and significance of the Enlightenment. Yet that is still not easy. In his dazzling and sympathetic account, written in the optimistic climate of the 1960s, Gay depicted the Enlightenment as a unity ('there was only one Enlightenment' [61: vol. 1, 3]), the work of a group who largely knew and admired each other, or at least were familiar with each others' works. They hailed from the major nations of Europe and British North America. There were the Frenchmen, Montesquieu, Voltaire, Diderot, d'Alembert, Turgot, Condorcet; the Britons, Locke, Hume and Gibbon; the Genevan, Rousseau; the German-born, d'Holbach, Kant and Herder; the American, Franklin. These constituted the hard core of what Gay called a 'family' or a 'little flock' of *philosophes*, flourishing from around the 1720s to the dawn of the new American Republic in the 1780s, when the French Revolution was on the horizon. And there were many others whose contributions were only slightly more peripheral or less influential: the pioneer psychologists, La Mettrie, Condillac and Helvétius; the codifier of utilitarianism, Jeremy Bentham; the Italian penologist and enemy of capital punishment, Beccaria; the systematizer of political economy, Adam Smith; those draughtsmen of American constitutional liberties, Jefferson, Adams, and Hamilton – and others besides.

Like the members of every close family, Gay cheerfully conceded, they had their disagreements. Yet he emphasized the cardinal points upon which they were essentially at one. They shared a general commitment to criticizing the injustices and exposing inefficiencies of the *ancien régime*; to emancipating mankind, through knowledge, education and science, from the chains of ignorance and error, superstition, theological dogma, and the dead hand of the clergy; to instilling a new mood of hope for a better future ('a recovery of nerve', Gay felicitously called it [61: vol. 2, ch. 1); and to practical action for creating greater prosperity, fairer laws, milder government, religious tolerance, intellectual freedom, expert administration, and not least, heightened individual self-awareness. Thanks to Gay's generous collective portrait of this 'party of humanity', the *philosophes* can no longer be dismissed as a bunch of pointy-headed intellectual *poseurs* [59].

Yet Gay's survey must be our point of departure in illuminating the Enlightenment, not the last word upon it. Many problems of interpretation remain outstanding, exposed by further digging in the archives or produced by new angles of vision. For one thing, there is the question of the relations between generals and rank-and-file. Gay's decision to devote his pages mainly to the 'great men' of the Enlightenment certainly honoured the towering reputation – 'notoriety' many would say – of the likes of Voltaire and Rousseau, men often condemned by reactionaries, as if they had almost single-handedly engineered the French Revolution. Gay's strategy enabled him to get under their skin, and to show they were complex human beings, whose ideas changed over time in response to experience – rather than just being names on the spines of books.

But more recent scholarship has looked away from these 'prize blooms' and paid more attention to the 'seedbed' of the Enlightenment. What sort of intellectual life, what groupings of writers and readers, made it possible for such giants to flourish? What conditions helped disseminate their teachings to wider audiences? Who continued their mission after their deaths? As well as a 'High Enlightenment', wasn't there also a 'Low Enlightenment'? Complementing the elite version, wasn't there also a 'popular' Enlightenment [47; 84]? These are issues taken up in Chapter 5.

The choice as to whether we see the Enlightenment principally as an elite movement, spearheaded by a small, illustrious band, or view it instead as a tide of opinion advancing upon a broad front, obviously colours our judgement of its impact. The smaller the leadership, the more readily the Enlightenment can be pictured primarily as a radical revolution of the mind, combatting the encrusted orthodoxies of the centuries with the new weapons of pantheism, Deism, atheism, republicanism, democracy, materialism, and so forth. We thrill to Voltaire thundering his magnificent cries of '*Écrasez l'infâme*' and '*Épater les bourgeois*' (outrage the bourgeoisie), making church and state tremble.

But another interpretation is possible; one in which the spotlight should fall less upon the embattled few than upon the swelling ranks of articulate and cultured men and women throughout Europe, those whom Daniel Roche has dubbed '*gens de culture*' (cultured people [130]); educated people at large, operating in the 'public sphere' [69] who preened themselves upon their own progressive opinions and 'polite' lifestyles, picking up a smattering or more of Voltaire and Co. – maybe just as a veneer, but sometimes as part of a genuinely new way of living.

Such a view would thus mean regarding 'Enlightenment' as a sea-change occurring *within* the *ancien régime*, rather than as the activities of a terrorist brigade bent on destroying it. So was the Enlightenment an intellectual vanguard movement? Or should it be seen as the common coinage of fashionable polite society? And in either case, did the Enlightenment actually transform the society it criticized? Or did it rather become transformed by it, and absorbed into it? In other words, did the establishment become enlightened, or the enlightenment become established? These issues will be examined below, in Chapters 6 and 7, and in the Conclusion.

Many other matters of interpretation hinge upon whether we see the Enlightenment as a 'militant tendency' operating amidst a hostile environment (ironically, rather as Gibbon portrayed the activities of the early Christians), or as a much wider ideology or *mentalité*. For one thing, we must broach the question of the practical impact of the Enlightenment in effecting change. The *philosophes*, as Gay has rightly emphasized, were contemptuous of dreamers with their heads in the clouds; they championed what Marxists were later to call 'praxis' (theoretically informed practical activity). When Diderot visited Russia at the behest of Catherine the Great, he explained to her at length that what her country above all needed were artisans and craftsmen [42: ch. 4]. Voltaire concluded his moral fable *Candide* (1759) by having the hero assert, '*il faut cultiver notre jardin*' (we must cultivate our garden – in other words, get on with things).*

But if (as Gay has argued) the flock of *philosophes* was 'small', and if most of them earned their living or won their fame as men of letters rather than as statesmen and politicians, can we realistically expect to find that they were Napoleons on the historical stage, possessing the power to change the very course of human affairs? Edward Gibbon, it is true, spent numerous years in the House of Commons as an MP, but, notoriously, he never made a single speech [119].

Of course, the chains of influence leading from attitudes to action are inevitably complex. Hence it may not matter that Voltaire never held office, because, we may point out, for many years he was in communication with the 'enlightened absolutist', Frederick II (the Great) of Prussia. Yet it seems that Frederick, far from heeding Voltaire, expected the *philosophe* to listen to him! Certainly, Frederick held advanced views (he was flagrantly irreligious), and he modernized the administration of his kingdom. Yet, despite the façade of sophisticated humanity, Frederick's Prussia – a militarized, war-hungry state indifferent to individual civil and political liberties – resembles a perversion of the true goals of the 'party of humanity' rather than their fulfilment [59; 60; 22].

What is at stake here is more than a matter of the influence of personalities, or questions of good or bad faith (did cynical enlightened absolutist rulers abuse naive *philosophes*?). Rather, it is a question of the *function*, no less than the *aims*, of Enlightenment ideals. The *philosophes* claimed that critical reason would prove emancipatory. Reason and science, they held, would make people more humane and happy. But certain scholars have recently been arguing that just the opposite occurred. When rulers and administrators heeded the promptings of 'reason', it was to increase their power and enhance their authority, in ways which often penalized the poor, weak, and inarticulate [55; 56]. Certain *philosophes*, such as the economists Quesnay and Mirabeau (known as the 'physiocrats'), claimed that free trade would increase prosperity. But when the French grain trade was finally deregulated, merchants profited and the poor suffered [104]. In a similar

---

*Though it is only the final ambiguity of a deeply ambiguous book – Voltaire's phrase may also have a largely private meaning: we must mind our own business.

fashion, the undermining of religion which *philosophes* encouraged led, some scholars claim, to the moral nihilism of the French Revolutionary Terror [45; 146].

In their *Dialectics of Enlightenment*, the German philosophers Horkheimer and Adorno have argued that it was thus no accident that 'reason' so often went hand-in-glove with 'absolutism' [80]. For reason and science, far from promoting liberty, encourage an absolutist cast of mind, by assuming an 'absolute' distinction between true and false, right and wrong, rather than a pluralist diversity of values. Along similar lines, the French thinker Michel Foucault has contended that Enlightenment principles and absolutist policy fused, in the name of rational administration, to promote cruel social policies. For instance, various kinds of social misfits – the old, the sick, beggars, petty criminals and the mad – were taken off the streets, lumped together as an 'unreasonable' social residue, and locked up in institutions. Here what purported to be 'enlightened' action was in reality repressive [55]. Postmodernists have accused the Enlightenment of promoting the absolutism of imperialist reason, while masquerading as tolerant and pluralist [49; 126]. Thus, it is not good enough simply to applaud enlightened intellectuals for attempting to tackle social problems; we must also assess the practical and ideological implications of their policies. It was one thing to deplore begging and the humiliating effects of dependence upon charity; it was another, however, to find effective solutions to the problems of which poverty was the symptom. Chapters 3 and 4 will attempt to evaluate the political and religious manifestos of the Enlightenment.

As is well known, Jean-Jacques Rousseau long ago contended that much of what other *philosophes* were commending in the name of reason, civilization and progress, would, in reality, render mankind only less free, less virtuous, and less happy [68]. The Genevan battled against Voltaire's unbelief and d'Holbach's materialism, both of which he saw as degrading. Yet Rousseau is always considered a leading light of the Enlightenment; and rightly so, for he was second to none in his hatred of the abominations of the *ancien régime*. Does not this suggest that the very label 'Enlightenment' may be more confusing than clarifying, if such a motley army of reformers could march under its banner?

The problem is real, but it is not unique to the Enlightenment. Without such labels, generalization would be impossible.

Sometimes we cannot in practice do without anachronistic labels – one's contemporaries did not pin upon themselves terms such as: reformers, radicals, reactionaries, and so forth. And the leading figures of what we call the 'Enlightenment' did, after all, see themselves as the bringers of light to the benighted.

There is, however, a particular problem with the movement we call the Enlightenment. This stems from the fact that, unlike certain agents in history, such as political parties or religious sects, it did not have a formal constitution, creed, programme, or party organization, nor was it committed to some explicit '-ology' or '-ism'. 'Dissenters' dissented from the Church of England, 'Chartists' endorsed the People's Charter. But there was no public charter of the Enlightenment, no party manifesto for the 'party of humanity'. Hence the Enlightenment is necessarily rather amorphous and diverse.

Yet, to admit this, does not mean to say that we would be better off abandoning talk of the Enlightenment altogether. Rather we should face up to this diversity. Thus, whereas, as quoted earlier, Peter Gay argued there was 'only one Enlightenment', Henry May has plausibly found four rather distinct types of Enlightenment in North America [103]. Taking May's hint, we might make a virtue of necessity and argue that such plurality, such lack of a sworn creed, may be amongst the distinctive and valuable features of eighteenth-century intellectual radicalism: perhaps its special strength – or maybe its fatal weakness, when contrasted to such a well-drilled body as the Jesuits.

What must not be masked is the fact that, for all its celebration of 'cosmopolitanism', the tone, priorities, and orientation of the Enlightenment differed from region to region and from kingdom to kingdom: a point which will be explored in Chapter 6 [139]. And this highly variegated quality of the Enlightenment must be borne in mind when we come, in the final chapter, to assess the true impact and significance of the 'siècle des lumières'. If we conclude that, despite the contentions of friends and enemies alike, the Enlightenment was not responsible for bringing about the French Revolution, would this be so decisive a verdict as if we concluded that the Communist Party failed, in some country, to spark a proletarian revolution? Enlightened activists perhaps never had such fixed targets in their sights.

In finally trying to assess the achievement of the Enlightenment, therefore, we would be mistaken to expect to find that a particular group of men effected a set of measures which amounted to 'progress'. Rather we should be judging whether habits of thinking, patterns of feeling, and styles of behaviour were modified, if not amongst the masses, at least among the many. Given that it was a movement aiming to open people's eyes, change their minds, and encourage them to think, we should expect the outcome to be diverse.

Indeed, it may be helpful to see the Enlightenment as precisely that point in European history when, benefiting from the rise of literacy, growing affluence, and the spread of publishing, the secular intelligentsia emerged as a relatively independent social force. Educated people were no longer standardly primarily the servants of the Crown or the mouthpieces of the Church. The pen may not have been mightier than the sword, yet Enlightenment words did prove dangerous weapons. Those making quills their arrows were not the grovelling mouthpieces of absolutist rulers, but freebooters, those intellectual bandits who have ensured the intellectual anarchy of 'free societies' ever since. These implications will be explored in greater detail in the concluding section of this book. But first it is necessary to examine the revolution in thought which the *philosophes* were trying to bring about.

2 The Goal: A Science of Man

Central to the aspirations of enlightened minds was the search for a true 'science of man'. Different thinkers had distinctive ideas of what this would involve. Hartley, La Mettrie and other 'materialists' (those who denied the independent existence of 'mind', 'spirit' or 'soul') hoped to develop a medico-scientific physiology of man understood as a delicate piece of machinery, or perhaps as just the most successful of the primates [150]. Some, such as Locke, Helvétius and Condillac, thought it was the mechanisms of man's thinking processes above all which needed to be investigated [71; 144]. Others, like the Italian Vico, believed man would best be understood by tracing the steps and stages of his emergence from some primitive condition or state of nature – which some envisaged as a golden age and others saw as a level of bestial savagery [134]. Still others, like Montesquieu and Hume, thought the key to a science of man lay in analysing the political and economic laws governing the interactions between the individual and society at large [104; 115; 31].

But, however great the differences of emphasis, there was widespread agreement that, in the words of Alexander Pope, the 'proper study of mankind is man'. Many sympathized with the aspiration of the Scottish philosopher David Hume, to create a science of politics and to be the 'Newton of the moral sciences' (or, as we would call them nowadays, the human and social sciences) [31; 57; 108]. Understanding why this quest for a science of man was both so attractive, yet also so fraught, will take us to the heart of the intellectual adventure of the Enlightenment.

Ever since Jakob Burckhardt's classic mid-nineteenth-century study, *The Civilization of the Renaissance in Italy* [32], it has often been said that it was the glory of early modern Italy to have discovered 'man' (as distinct from the son of Adam, the Christian pilgrim), and in particular to have developed the idea of human 'individuality'. There is a truth in this, and in the hands of daring spirits, such as Montaigne in sixteenth-century France, who posed the sceptical question, *'que sçais je?'* (what do I know?), Renaissance Humanism could lead to searching introspection into the human condition. Shakespeare has Hamlet muse, 'what a piece of work is man' [61: vol. 1, ch. 5].

Yet the 'man' the Renaissance discovered was typically also a somewhat conventional figure. He was still the being initially created, in the Garden of Eden, whole and perfect by the Christian God in His own image – in that sense, the humanist philosophy of the Renaissance was every bit as Christian as the faith of Luther or of the Council of Trent. Renaissance man, ostentatiously portrayed by artists as the well-proportioned, handsome nude, or the geometrically regular 'Vitruvian man', could still be represented as the microcosmic analogue of the macrocosm at large (the little world of man as an emblem of the great world of the universe). Practically all sixteenth-century thinkers, Copernicus excepted, still believed in the 'homocentric' (man-centred) and 'geocentric' (Earth-centred) cosmos first advanced by classical Greek science, with man as the measure of the divinely-created system of the universe. Likewise, most Renaissance scholars felt confident enough in tracing human history back, through a continuous pedigree, to Abraham, to Noah, and ultimately to Adam, the first human. Man thus retained his divinely fixed place in time and space.

Admittedly, the new Renaissance adulation for things Greek and Roman disturbed those evangelical churchmen who preached that Christ had died to redeem mankind from sin and the errors of paganism. But the broad effect of Renaissance Humanism's 'anticomania' (love of Antiquity) lay in consolidating a reassuringly harmonious vision of human nature and destiny. Moralists believed that from Classical poets, philosophers, moralists, historians and statesmen – above all, from Xenophon, Seneca, Cicero and Livy – models of virtue could be derived which the truly civilized man could pursue, in harmony with the Christian's progress towards spirituality and salvation.

The Renaissance thus emphasized dual but mutually consonant aspirations for man. It restored Classical learning, and thereby recovered a this-worldly model for social and political living. But it also integrated these noble ideals of Antiquity with the purified truths of Christianity as spelt out in Scripture and authorized by the Church. These twin goals, uniting the good man and the good Christian, commanded widespread acceptance for well over a century.

Very slowly, however, they came apart at the seams. For one thing, the ferocious religious and dynastic struggles racking Europe from the Reformation through to the close of the Thirty Years War (1648) inevitably challenged the optimistic Renaissance faith that man was a noble being destined to fulfil himself through engaging in the public life of the commonwealth: Machiavelli's cynical and pessimistic view of man made itself felt. For another, with the progress of historical scholarship, a new sense of the past emerged, which finally laid dramatically bare the glaring divide between the 'old world' of Graeco-Roman Antiquity and the 'new world' of statecraft and diplomacy, of guns and the printing press [33]. Moreover, genuinely 'new worlds' were being discovered, above all America, unknown to the Ancients, presenting scenes of exotic, heathen and savage life that challenged Renaissance intellectuals' cosy assumption that Florence was the modern Athens, and the Holy Roman Empire was the successor to Rome itself.

What is more, the seventeenth century was to prove far more intellectually corrosive than the sixteenth. The brilliant 'new sciences' of astronomy, cosmology and physics, pioneered by Kepler, Galileo, Descartes and their successors, destroyed the old harmonies of an anthropocentric (man-centred) universe, that small closed world focused upon man himself, which both Greek science and the Bible had endorsed. Copernican astronomy, assimilated in the seventeenth century thanks to a succession of geniuses from Kepler to Newton, displaced the Earth, and man upon it, from being the centre of the universe. It ended up a tiny, insignificant planet, nowhere in particular in that dauntingly infinite universe (now visible through the newly developed telescope) whose immense spaces so frightened Pascal [40; 74; 121].

The new 'mechanical philosophy', espoused by 'atomistic' scientists who claimed that Nature comprised nothing but particles of matter governed by universal laws whose actions could be expressed mathematically, was, of course, a tremendous triumph of investigation and conceptualization. But it left what had always been cast as 'Mother Nature' dead and impersonal. The French

13

philosopher and scientist René Descartes, moreover, contended that all living creatures, man alone excepted, were merely machines or automata, lacking even consciousness. The possibility inevitably arose that man himself might be just another machine – one, however, prone to vanity and self-delusion [84].

Early in the seventeenth century, the Metaphysical poet John Donne declared, 'And new philosophie calls all in doubt'. It would be quite wrong to imply that, faced with the discoveries of the 'new science', all thinkers doubted and despaired. But, in the light of this radical transformation of theories of Nature, many believed that received ideas about the history, nature, and destiny of man had themselves to be re-examined.

And a further, unsettling element became more prominent in the second half of the seventeenth century. Ever since the Reformation and Counter-Reformation, a ferocious polemical war had been waged between Protestant and Catholic Biblical scholars and theologians over the fundamentals of faith. Central to these battles were rival contentions about who, where, and what precisely was the True Church; whence its authority was derived; whether every syllable of the Scriptures was inspired and literally true, and so forth.

Such wrangling, often acrimonious and unedifying, inevitably, in the eyes of some free spirits and inquiring minds, sapped the moral authority of the churches. Worse, it drove acute and honest scholars face-to-face with the profound questions of man's history and destiny, which close scrutiny of the Bible forced to be asked but did not (it now seemed) readily answer. Could the world really be only 6000 years old, as the Bible stated? Was Adam truly the first man? Did a serpent really hold a conversation with Adam and Eve in the Garden of Eden? Could a just and benevolent God really have exterminated the whole of the human race, save only Noah's family, at the Deluge? In any case, where had the water for Noah's Flood come from? Where had it gone? Had the Flood been a miracle? Or was it – and perhaps many other Biblical 'miracles' besides – to be explained as an 'effect' of the regular laws of nature as now, at last, understood by modern science? Had the Sun literally stood still for Joshua at the siege of Jericho? – and so on. Questionings of this kind uncovered hundreds of issues – historical, moral, scientific and theological – which posed pressing difficulties of fact and faith that Christians needed to settle. The authority of revealed religion was being questioned. Some better path to true knowledge had to be sought.

The *Dictionnaire* (1697) of the unorthodox Huguenot Pierre Bayle, who had sought refuge in Holland from Louis XIV, gave great prominence to such doubts and dilemmas. Bayle also pinpointed the childish absurdities of pagan worship, in a manner that could be taken as a veiled attack upon Christianity itself. Scholars disagree whether Bayle was, at heart, a 'fideist', that is a believer who thought it the Christian's duty to assent to the authority of faith, as a means of overcoming rational doubt; or whether he was, rather, a sceptic, taking delight in spreading doubt and confusion. He was certainly adroit in covering his tracks [74; 75; 84; 134].

From the latter part of the seventeenth century, many of Europe's greatest minds came to the conclusion that to understand the true history and destiny of the human race, neither unquestioning faith in the Bible, nor automatic reliance on the authority of the Greek and Roman thinkers (the 'Ancients') would any longer suffice. Man's nature was not properly known; it must become the subject of inquiry. And the proper engine of such an investigation must be that 'scientific method' which natural scientists (the 'Moderns') had pioneered so successfully in the fields of astronomy, physics and anatomy [61: vol. 2, ch. 3; 70].

Systematic doubt, as advocated by Descartes, experimentation, reliance upon first-hand experience rather than second-hand authority, and confidence in the regular order of Nature – these procedures would reveal the laws of man's existence as a conscious being in society, much as they had demonstrated how gravity, as Newton proved, governed the motions of the planets in the solar system. This kind of analogy with natural science was precisely what Hume had in mind when he spoke of becoming the 'Newton of the moral sciences' [28; 31; 115]. For the new 'social scientists' of the Enlightenment, the old 'truths', expounded by Christianity and the pagan classics, now became open to question; in this respect at least, the 'Moderns' had surpassed the 'Ancients' in what was often dubbed the 'Battle of the Books' (the debate as to whether modern minds truly excelled the Greeks) [90].

Enthusiasts for Enlightenment were thus fired by Francis Bacon's conviction that the methods of natural science would launch the 'advancement of learning'; such newly acquired knowledge would

15

lead to power, and thereby, in Bacon's phrase, to 'the effecting of all things possible'. As Voltaire emphasized in his *Lettres philosophiques* (1733), Newton's achievement truly demonstrated that science was the key to human progress [73]. Or, in Alexander Pope's couplet,

> Nature, and Nature's Laws lay hid in Night:
> God said, *Let Newton be! and All was Light.*

If the Roman Catholic Church chose to pronounce Copernicanism heresy, and to persecute Galileo, that merely proved that truth always had its enemies. Yet truth was great and would prevail.

The French historian Paul Hazard termed this late seventeenth-century time of ferment and unsettlement the 'crisis' of the European mind [74; 75; *cf.* 88]. Enlightened minds believed that such a 'crisis' was to be overcome by the execution of a programme for the scientific understanding of man. One favourite attempt along these lines lay in the construction of a 'natural history of man' to replace the traditional 'sacred history' of the Old Testament. Many *philosophes* tried to develop, empirically, imaginatively or systematically, such a historical or anthropological vision, tracing the emergence of European man out of the state of 'savagery' which was assumed to have been his primeval origin, and which could be inferred from the 'primitive' condition of the tribes explorers were beginning to discover in darkest Africa, America and, eventually, Australia [31].

To put such aboriginal peoples' capacity for progress to the test of science, natives were sometimes transported to Paris or London, and then exposed to the laboratory of polite society. A Polynesian, Omai, was brought back from the newly discovered Tahiti. Similar experiments were performed upon *enfants sauvages*, feral children found running wild in the woods of Europe. The Enlightenment faith in future progress, in the secular perfectibility of man, as proclaimed by Herder and Condorcet, and by such Scottish philosophers as Ferguson and Millar, hinged upon the assumption that much of mankind had already risen from 'savagery' to 'civilization', from 'rudeness' to 'refinement', or from the savage to the Scotsman [31; 19; 79].

Such supposition about the human capacity for progress would, of course, have been unthinkable without belief in the extraordinary plasticity of man's faculties, and a generous confidence in the species' capacity for learning, change, adaptation and improvement. Fundamentalist Christian theologies, Catholic and Protestant alike, had traditionally characterized man as irremediably flawed by the 'original sin' of the 'Fall': without faith, or the sacraments of the Church, all man did was evil. The philosophical pessimists of Classical Antiquity had likewise seen man as inevitably engaged in constant civil war with himself, his nobler faculty of reason being all too easily overwhelmed by rebellious appetites and passions. Hence, thought the Stoics, a certain aloof detachment from his baser self was the best state man might hope to achieve [154].

The new Enlightenment approaches to human nature, by contrast, dismissed the idea of innate sinfulness as unscientific and without foundation, arguing instead that such passions as love, desire, pride and ambition were not inevitably evil or destructive; properly channelled, they could serve as aids to human advancement [45]. In Bernard Mandeville's paradoxical formula, 'private vices' (like vanity or greed) could prove 'public benefits' (for instance, by encouraging consumption and thereby providing employment) [81]. Such Enlightenment thinkers as Helvétius in France, and the pioneer utilitarian Jeremy Bentham in England, developed a psychological approach. Replacing the old moralizing vision of man as a rational being threatened by brutish appetites, they newly envisaged man as a creature sensibly programmed by nature to seek pleasure and to avoid pain. The true end of enlightened social policy ought therefore to be to encourage enlightened self-interest to realize the 'greatest happiness of the greatest number' [51; 71; 102; 144].

Traditional preachers would have denounced such advocacy of the 'pleasure principle' as sinful, brutish hedonism. But a new breed of political economists, notably the Scottish Adam Smith in his *Wealth of Nations* (1776), contended that the selfish behaviour of individual producers and consumers, if pursued in accordance with the competitive laws of the market, would result in the common good – thanks, in part, to the help of the 'invisible hand' of Providence [31; 38; 71; 79; 114; 133]. Likewise, such legal reformers as the Italian Beccaria argued that a truly scientific jurisprudence needed to be built upon the assumption of a psychology of rational selfishness: the pains of punishment must be precisely calculated to outweigh the pleasures of crime [152].

If mankind were to be progressive, the species had to be capable of change; above all, of adapting to new environments. Not surprisingly, therefore, Enlightenment psychologists were preoccupied with the learning process, and held out great hopes for education. The history of the race, suggested many thinkers, following Locke's fundamental *Essay Concerning Human Understanding* (1690) and his *Some Thoughts Concerning Education* (1693), could be seen as like the education of the individual infant writ large [165]. Followers of Locke believed that preachers were wrong to judge that man was born sinful, and that Plato had been equally mistaken in claiming that people were born ready kitted out with 'innate ideas' (for instance, those of right and wrong). Rather, the human mind began as a *'tabula rasa'*, a clean slate or a 'blank sheet of paper'. It then continually absorbed data through the five senses, storing this information and shaping it into 'ideas', which were destined to become our empirical knowledge of the world and our moral values. Man's nature, capacities and knowledge were thus entirely the product of learning from experience, through a process involving the association of ideas (the building of complex ideas out of simple units). Man was thus the child of his environment; but in turn he acquired the capacity to transform those same surroundings.

Engaged thus in a constant dialectical interplay with his fellows and environment, man was ever evolving to meet the challenges of a world he was continually modifying. Hence it followed for such admirers of Locke as Condillac and Helvétius that man was his own maker, and that his self-developing potential knew no hard-and-fast bounds. Towards the close of the eighteenth century, Condorcet wrote his *Esquisse d'un tableau historique de l'esprit humain* [*Sketch for a Historical Picture of the Human Mind*] (1794), which charted, in terms ever more rapturous as the future was approached, all the stages of the progress of the human mind – past, present and to come. Condorcet (who was driven to commit suicide in the French Revolution) boldly suggested that man, thanks to his capacity for 'perfectibility', would soon overcome want, weakness, disease and even death itself [19; 43: ch. 6]. The English anarchist William Godwin thought similarly. Both the French naturalist Lamarck and his English contemporary, the doctor and scientist Erasmus Darwin – Charles Darwin's grandfather – outlined the first biological theory of evolution, which presupposed, in their different ways, just such a capacity of creatures to learn, change, adapt and pass on their acquired characteristics to their offspring [96].

As it moves into the twenty-first century, Western civilization still subscribes to – or, rather, some would say, remains imprisoned within – this secular vision of the limitless human drive towards economic growth, scientific innovation, and progress, which the Enlightenment envisaged. Today's social sciences – sociology, economics, psychology, anthropology – have all emerged from seeds sown in the Enlightenment [18]. Prime ministers such as Mrs Thatcher have recently appealed to the teachings of Adam Smith, to justify their faith in market forces and the capacity of the pursuit of profit to guarantee the general good.

In view of this, we must consider the ambiguities of the science of man as forged in the eighteenth century, and note the complexities of its legacy. The *philosophes* claimed that they had dynamited obsolete religious 'myths' about man, and his place, under God, in Nature, replacing them with true scientific knowledge, objectively grounded upon facts. Many historians, including Gay, praise them for thus breaking with 'mythopoeic' thinking, and advancing 'from myth to reason' [61: vol. 1, ch. 2].

But it might be better to say that what the *philosophes* essentially did was to replace a *Christian* myth with a *scientific* myth – one more appropriate for an age of technology and industrialization. At bottom, it has been noted, the two myths have remarkably similar configurations. As Carl Becker contended in his wittily titled *Heavenly City of the Eighteenth-Century Philosophers*, the idea of the state of nature, as developed by *philosophe* speculative history, bears an uncanny resemblance to the Garden of Eden and the Fall as envisaged by Christian theology. Similarly, the Enlightenment idea of indefinite future progress can be seen as the secularization of the doctrine of Heaven. Far from being cast-iron 'facts', the notions of the noble savage and of progress are just as speculative, symbolic, and dependent upon preconceptions – faith even, one might argue – as the Christian formulations they succeeded [21; 154].

To suggest that the Enlightenment offered, not science in place of myth, but new myths for old, is not to debunk it. But it means that we must not take Enlightenment claims at face value, but treat them as highly effective propaganda. Consider, for instance, the development of economics. In his *Wealth of Nations* (1776), Adam Smith berated governments for their traditional 'mercantilist' and 'protectionist' policies, which (he argued) hamstrung trade for the fiscalist purposes of raising revenue. Smith further attacked the traditional belief that war was the route to wealth; accused vested interests of supporting monopolies contrary to the public interest; and argued that, properly understood, market mechanisms would, in the long-term, prove beneficial to all. In the light of such claims, we may understand why Gay concludes that Smithian, *laissez-faire* economics were more 'humane' and 'scientific' than the systems they challenged [61: vol. 1].

But it is also important not to forget that Smithian (or 'classical') economics provided an apologetics for capitalism in an age of industrialization, not least through its recommendations for the deregulation of labour (euphemistically called 'free labour'). Smith himself was frank enough to admit that the extreme division of labour required by modern manufacturing – his prime example was pin-making – reduced the worker to a 'hand', a mentally-stunted, alienated, slave-like machine. But he was not 'humane' enough to suggest a remedy. Classical economics' theory of the laws of profit and loss and the 'iron law of wages' precluded such 'interference' with market mechanisms (all interference with competition, they claimed, only encouraged inefficiency). *Laissez-faire* economics thus endorsed an *inhumane* system in the name of the 'natural laws' of market forces – laws which, the politician Edmund Burke proclaimed, were sacred because they were the 'laws of God'.

The new social sciences developed by the *philosophes* were highly critical of Christian conceptions of divinely appointed government, and of feudal hierarchy and subordination. But (with a few exceptions, such as Rousseau [42; 67; 116]), they did not provide anything like such a searching critique of commercial society, with its sanctification of private property and individual interests. In many ways, the Enlightenment hymn to 'progress' turned a blind eye to the equally biting inequalities and oppressions of the new commercial and industrial order: after all, wasn't everything getting better [97]? It is no accident that Blake, the Romantic visionary, so passionate in his denunciation of 'dark, satanic mills', should have condemned such leading *philosophes* as Bacon and Locke, Newton and Voltaire, as the evil geniuses behind that system. Maureen McNeil has plausibly argued that Erasmus Darwin, doctor, educationalist and scientist – overall the leading *philosophe* of late eighteenth-century England – was also the most articulate enthusiast for the values of the new industrial society [96].

3 The Politics of Enlightenment

The political ideas of the *philosophes* have always had their critics. Such opponents of the French Revolution as Edmund Burke and the Abbé Barruel portrayed them as immature rationalists, whose *a priori* and irresponsible sloganizing in favour of abstract liberty, the general will, and the rights of the people helped to topple the old order, only to produce first anarchy, and then a new despotism, in its place. Above all, critics complained, in politics the *philosophes* lacked that quality they pretended to value most: experience.

The charge is not without superficial plausibility. Whereas the conservative Burke spent a lifetime in Parliament, held office and tasted power, most *philosophes* did not get beyond being mere parlour policy-makers. It was hardly their fault, however. Louis XV was prepared in 1745 to appoint Voltaire 'historiographer royal'; he was, after all, the most eminent historian, playwright and poet of his generation [30]. But the monarch was hardly likely to appoint such a scabrous critic a minister of the Crown [60].

The political preferences of certain *philosophes* can be made to seem mighty unrealistic, naive, or even nightmarish. In his *Contrat social*, Rousseau praised small, poor republics as the nurseries of public virtue; but his adulation for ancient Sparta and the early Roman republic were at best only obliquely relevant in the Europe of the mid-eighteenth century, where the few remaining city states, such as Geneva and Venice, were oligarchic and opulent. In the opinion of certain recent historians, for instance J. L. Talmon, judging Rousseau in the light of the twentieth-century experience of fascism, his call for a heroic 'legislator' to act as a national regenerator was at best simple-minded and at worst sinister (though we may see such judgements as anachronistic). And what are we to make of the logic of his promise, or threat, to 'force men to be free'? [42; 43: ch. 3; 68]. Other thinkers were, arguably, no less unrealistic. In his *Inquiry Concerning Political Justice* (1793), William Godwin advanced such a stridently individualist brand of anarchism, that he not only condemned marriage but also denounced orchestras and the theatre, on the grounds that they enforced conformity and compromised individuality.

The *philosophes* have also sometimes been accused of being politically rather unprincipled. The self-same Godwin who passionately denounced marriage as an infringement of individual liberty (it was a kind of legal prostitution) not only subsequently wed the freethinking feminist, Mary Wollstonecraft, but insisted that the poet Shelley in his turn should marry their daughter Mary, rather than merely cohabit with her. Voltaire and Diderot, for their part, flirted with, and flattered, the leading absolutists of Europe, Frederick the Great of Prussia and Catherine the Great of Russia. Their patronage, support and protection were doubtless useful; but the relative silence of these intellectuals, when confronted with the internally oppressive and externally bellicose policies pursued by both autocrats, leaves many questions to be answered [59; 89; 165].

Eminent French *philosophes* boldly denounced the evils of the old regime: but can it be said that they matched their eloquence with real political action? None seriously went about organizing political resistance, or issued calls to arms. Was this perhaps because, at bottom, they mainly felt comfortable enough with the *status quo*? Voltaire and Diderot both suffered short spells of political detention; but, that aside, they were able to proceed with their subversive labours without too much jeopardy to their personal liberty, and were lionized by the literary salons. Contrast the bloody fates in the sixteenth and seventeenth centuries of thousands of heretics and of freethinkers such as Bruno, Campanella, and even, although to a lesser degree, of Galileo; and recall the persecution and outlawing of agitators in nineteenth-century Russia or Austria. How committed, therefore, were the *philosophes* to destroying the *ancien régime*? Might they rather be dismissed as noisy political lightweights?

In assessing their significance as political thinkers and activists, many things must first be taken into account. For one thing, ever since the American and French Revolutions, a programme of political goals has established itself definitively in the Western world. We all now believe in government of the people, by the people, for the people. We believe in universal suffrage. We believe that democracy safeguards liberty; we believe in parliament, in elections, in representative government, in the party system. For better or worse, these have become the sacred cows of Western 'democracy'. Why then didn't the *philosophes* champion them?

The answer is that there is not the slightest reason why an enlightened intellectual in 1700 or 1750 should automatically have endorsed any of these principles, still less the whole package. Parliaments (in France, the *parlements*) had traditionally operated as the bastions of aristocratic vested interests, and political parties were universally associated with self-serving factionalism. Direct democracy was a system of government which had come in, and had gone out, with the ancient Greeks. And, as Rousseau knew as well as any radical English journalist, representative government was a recipe for gerrymandering and corruption [52; 116].

Above all, what possible grounds could the *philosophes* have had for vesting political trust in the wisdom of the people at large? Almost everywhere in Europe, the bulk of the population consisted of illiterate peasants, labourers, and (east of the Elbe) even serfs – all, to elitist eyes, hopelessly ignorant, backward, and superstitious, browbeaten by custom into an unthinking deference to Throne and Altar. The likes of Voltaire depicted the peasantry as barely distinguishable from the beasts of the field. Their point in making such unflattering comparisons was to criticize a system which reduced humans to the level of brutes; but such comments betray a mentality for which the true question was not popular participation in government – that did not seem a high priority – but whether the people were to be ruled wisely or incompetently [60; 100; 113].

The question of the right or legitimate type of government – who precisely should legally, or expediently, hold power – was raised in its most comprehensive form by Montesquieu in *L'Esprit des lois* (1748) [142]. *The Spirit of the Laws* identified three chief types of polity. First there were republics: Montesquieu felt a strong leaning towards republican government, believing that its participatory form preserved the liberty and enhanced the virtue

24

of those actively engaged in political life within them. Republics had flourished in Antiquity, and Montesquieu ruefully concluded they were essentially a thing of the past [43: ch. 1].

Next there were monarchies, clearly a viable form of government in the modern world. They derived great stability from the hierarchical gradation of ranks they bolstered, which conferred a well-defined place upon nobles, gentlemen and ecclesiastics, and also from the sense of 'honour' which every member of a group attached to his rank. Monarchy was the desirable form of one-man rule. Its perversion, thirdly, was despotism, in which the ruler levelled all rightful distinctions between his subjects, and governed by fear. It was Montesquieu's anxiety – derived from consideration of the ambitions of Louis XIV – that the French crown was aspiring to change itself from 'monarchy' into 'despotism'. Hence his own writings, by way of counterweight, celebrated the political role of the traditional aristocracy, the provincial *parlements*, and even the Church, in the hope that these could serve as buffers, checks and balances, to prevent the emergence of Bourbon 'despotism'.

Montesquieu's analysis laid bare the dilemmas of the day in a particularly bleak form. In modern, big-state dynastic politics, republican government was evidently obsolescent. Monarchy, however, was gravitating towards despotism. Hence the preservation of liberty would probably require the support of the most reactionary estates of the realm (Montesquieu had already devastatingly mocked the pretensions and privileges of the nobility in his *Persian Letters*).

In any case, the politics of the *parlements* also had their dangers. Frenchmen were to find to their cost that whenever Louis XV or his successor attempted much-needed economic reforms or budgetary rationalizations, the *parlements* and nobility possessed the power to stymie change. More pessimistically yet, Rousseau was to argue that salvation did not lie in devising more sophisticated forms of constitutional arrangements to prevent the abuse of power and privilege. For the very fibre of modern society itself was utterly 'corrupt', alienating man from man in ways that sapped liberty, destroyed virtue, and caused decay.

The problems of who should govern proved perplexing; the *philosophes* on the whole found it more constructive to formulate advice about what rulers should *do*. They did not envisage that government ought to be a simple matter of 'legitimism' and hereditary succession, the maintenance of the *status quo*, the defence of existing property rights and privileges. They wanted administrations which would achieve improvements, by promoting peace, prosperity, justice and welfare within civil society. As part of this goal, they naturally deplored all undue interference with the personal affairs of subjects [59]. Liberty of thought and expression, freedom to publish, religious toleration and the right of minorities to worship – all these were the elementary requirements of social beings [66]. Even the Prussian Kant, a temperamental conservative who distrusted the idea of the right of popular participation in government, argued that it would be degrading, both to government and to the governed, to deny basic civil rights, for that would be tantamount to treating adults as children. Voltaire extolled the virtues of civil and religious liberty, English style, by picturing in his *Lettres philosophiques* (1733) the scene at the London Stock Exchange. There Anglicans, Dissenters and Catholics, Jews and Mohammedans were all permitted to deal on equal terms. Freedom of trade went with freedom of religion, bringing peace and prosperity [60].

Philosophes deplored what they saw as the erosion of freedom throughout most of Europe. 'Man was born free,' Rousseau celebratedly opened his *Social Contract*, 'but everywhere he is in chains.' But the Continental *philosophes* of the 'High Enlightenment' never made their prime demand the maximization of personal freedom and the reciprocal attenuation of the state, in the manner of later English *laissez-faire* liberalism. For one thing, a strong executive would be needed to maintain the freedom of subjects against the encroachments of the Church and the privileges of the nobles. Physiocrats like Quesnay championed an economic policy of free trade, but recognized that only a determined, *dirigiste* administration would prove capable of upholding market freedoms against entrenched vested interests [104]. No Continental thinkers were attracted to the ideal of the 'nightwatchman' state so beloved of English radicals. Even Tom Paine, whose *Rights of Man* (1791) inveighed so vehemently against tyrannical oppression, nevertheless considered that a lawfully constituted popular government, duly elected by the people, ought to pursue constructive welfare policies (e.g., introducing old age pensions and child benefits) [39].

It was the thinkers of Germanic and central Europe above all who looked to powerful, enlightened rulers to preside over a 'well-policed' state [89; 127]. By this was meant a regime in which an efficient, professional career bureaucracy comprehensively regulated civic life, trade, occupations, morals and health, often down to quite minute details. Laws were to be passed, for example, giving encouragement to earlier marriage, thereby boosting population, increasing the workforce, stimulating the economy, and extending the tax base, and with it the potential military strength of the realm. Such leading advocates of systematic rational government as Justi and Sonnenfels argued that *Cameralwissenschaft* (the science and practice of administration) would simultaneously serve the ruler (by increasing revenue and strengthening public order) while also improving the lot of the people [26]. In the German principalities, medical practice, for example, was exhaustively regulated by the state, supposedly for the public good [91].

A similar disposition to regulation, or at least to giving priority to a centrally-organized 'general good', is also prominent in the more radical French theories of the mid-century. The utilitarian Helvétius saw human beings as essentially identical. All were thus equally malleable, capable of being profoundly conditioned by education and environment. The wise ruler could thus ensure the happiness of all [144]. Other thinkers such as Mably and Morelly contended that the abolition of sectarian privilege and private property, and the subordination of individual interests to the common good, would give rise to a new breed of virtuous citizens. Rousseau similarly dreamed that in a properly constituted society, human nature itself would be regenerated – or, at least, its degeneration would be halted [28: ch. 3, 5].

The thinkers of the early Enlightenment were preoccupied with finding ways to check the spread of despotism. Locke championed constitutional government, arguing that all legitimate authority was circumscribed by the laws of nature and derived from the consent of the governed – the Whig philosopher, however, was no democrat, and he certainly never envisaged political rights for women [97; 99]. Montesquieu used historical arguments to confirm constitutional prerogatives. By mid-century, attention had shifted, I have been suggesting, to questions of the ends and uses of political power. What kind of a state

27

would produce virtuous men? What sorts of policies would expedite trade or make for a healthy populace? Political programmes thus grew more positive, but they also ran the risk of degenerating into a proliferation of wish lists or even utopian fantasies [151].

That is why the revolt of the American Colonies against Britain proved utterly decisive for Europe. The Declaration of Independence (1776) demonstrated that, at least in the new world, virtuous action in defence of liberty was still possible. The War of Independence proved the military calibre of a citizen army, defending freedom against the British 'tyrants'. The American Constitution confirmed that republican modes of government could function in the modern world, and set an example of the role of the people as the sovereign political body [162; 61: vol. 2, 'Finale'].

At last, in the American Constitution, a political model existed in which the defence of liberty did not depend, as with Montesquieu, upon reactionary social groups. The American experiment – and it seemed to work – appeared to prove that power did not automatically corrupt, provided it came from the people and was carefully constitutionally regulated. The new republic beckoned to old Europe to go and do likewise. In the 1780s, such *philosophes* as Condorcet, who had heretofore trusted to reason and expertise to secure improvement, abandoned their earlier technocratic bias, and began to discover a new political virtue in the people [19; 117].

The ambiguity of the question of who exactly constituted, or could speak for, the people, did not become crystal-clear until after 1789. The French Revolution of course created a ferment of political theories and schemes. Following the execution of Louis XVI, successive regimes found no stable and successful way of reconciling the twin demands of popular government and efficient government. Yet that is to say no more than that the Revolution itself failed to resolve the riddle that had taxed the Enlightenment's best brains.

4 Reforming Religion by Reason

The waters of politics proved treacherous for the *philosophes* to navigate. They were unsure whether they wanted power themselves, or would prefer to exercise the prerogative of the pundit in criticizing the prince. They were more sure of their ground, however, with religion. No religion that lay beyond their own personal jurisdiction was acceptable to them. Indeed, for many Enlightenment minds, religion was unacceptable in any established ecclesiastical shape or form.

Key Enlightenment thinkers were pained and angered by the intellectual and institutional manifestations of religion they witnessed all around them. Many leading *philosophes* – above all those from France and England – made bitter and mocking onslaughts upon the absurdity of Christian theology, the power-crazed corruption of the churches (above all, the Vatican), and the pestilential power still exerted by blind credulity over people's lives. For some, notably Voltaire, Diderot and d'Holbach, the emancipation of mankind from religious tyranny had to be the first blow struck in a general politics of emancipation, because the individual possessed by a false faith could not be in possession of himself [34; 98].

Voltaire made his crusade – first against Popes, Jesuits and priests, and finally against the Christian God – the climax of his career [60]. David Hume deployed his sceptical philosophy of experience to destroy the traditional argument that God's existence and qualities could be demonstrated from Creation. Our grasp of the relations between cause and effect, he argued, depended upon experience of multiple examples; but there was only one single universe, or 'effect'; hence we were in no position to assess its 'cause' or creator. For Hume, furthermore, the idea that a reasonable man could believe in miracles was a contradiction in terms. In his *Decline and Fall of the Roman Empire* (1776–88), Edward Gibbon contended that Christianity had been no less responsible than the barbarian invaders for laying low that majestic edifice of civilization, and thus ushering in a millennium of darkness [118; 119: ch. 5].

Trenchard, Shaftesbury and the Deists in England [36], and Fontenelle, Boulanger and others in France, argued that the popular belief in the existence of wrathful gods in the sky should be interpreted primarily as the sick response of the savage mind, terrified of the unknown and powerless to cope with the forces of Nature. What primitive people feared, they turned into objects of grovelling worship. They had, so such *philosophes* argued, invented magic and sacrifices to placate these fabulous deities. Their readers understood, of course, that their account applied not only to the tribes of Africa or North America. For their real target was Christianity, with its magical sacraments and repeated sacrifice of Christ in the Eucharist [98].

In his *Lettres persanes* (1721), Montesquieu, speaking through the mouthpiece of a Persian traveller, called the Pope a magician. Along similar lines, the Baron d'Holbach exposed religion as an infantile precursor to science. The primitive mind, he claimed, believed in souls and angels, devils and witches, and other childish fantasies [62]. Mature reason, by contrast, proved that none of these existed. All there was, was Nature, and Nature was simply a material system of physical objects governed by the regular action of the immutable laws of science [98].

Above all, Voltaire was the anti-Christ of the Enlightenment, battling throughout his career against the demons of false religion. Ever ready with an anti-clerical jest, his early campaigns were waged largely in the cause of religious tolerance (he particularly admired the peaceful English Quakers), and were directed against the Church Militant, and its more outlandish beliefs and practices [66]. Over the decades, his antipathy grew fiercer, and he directed his ferocious moral passion against the evils Christianity had perpetrated through wars of religion, burning heretics, executing so-called 'witches', etc.

In place of the Christian creed and church, Voltaire hoped to install what Gay has called 'modern paganism' [61: vol. 2, ch. 7].

This would take the form of 'natural religion', a non-dogmatic belief in the existence of a rational, benevolent God, to be regarded as the author of the Newtonian universe, and revered as the guarantor of justice and morality amongst men. Take away such a God, mused Voltaire, and why should not people be wicked with impunity? (That was why, as he quipped, if God did not exist, it would be necessary to invent Him.)

But even in this kind of 'natural religion', Voltaire was to suffer a loss of faith. In particular, the Lisbon earthquake of 1755, that cosmic massacre of the innocents, shook his confidence in the Benign Intelligence he hoped he had seen behind Nature. In his last decades, Voltaire railed with restless, relentless ferocity against all religion, as though God (who after all did not exist) had done him some personal injury. His reiterated rally-call, '*Écrasez l'infâme*', was extended beyond the Pope, beyond the organized churches, to practically all manifestations of religiosity whatsoever. Voltaire ended up, perhaps, an atheist.

We shall have to ask why so many of the leading *philosophes* – men who prided themselves upon their own sophisticated tolerance, their man-of-the-world acceptance of human foibles – grew apoplectic, or descended into sarcasm and smut, when confronted with priests and creeds. After all, the high noon of torture and heretic- and witch-burnings in the name of the true faith, of inquisition and crusading, was a thing of the past. Holy wars had had their day, and the Christian churches of eighteenth-century Europe had grown relatively torpid and tolerant: no small number of *philosophes* – Raynal and Mably, for example – were themselves *abbés*, while the Archbishop of Canterbury's wife happily played whist with the arch-infidel Gibbon [66; 119].

It is important, however, to insist upon the range and complexity of Enlightenment attitudes towards faith. Very few intellectuals wanted to replace religion with out-and-out unbelief. For one thing, most believed that science and philosophy, though casting doubt upon the existence of the specifically Christian, Biblical, anthropomorphic 'God of miracles', nevertheless pointed to some sort of presiding Deity, a supernatural Creator, Designer, and Mind. Even the sceptic Hume thought that it was improbable that the orderliness of the cosmos had come about by accident or chance, from the purely random movements of material atoms: in our idiom, such an outcome was as unlikely as a chimpanzee bashing typewriter keys and producing Shakespeare.

Furthermore, many *philosophes*, while hostile to 'priestcraft' and high church ecclesiastical pomp, nevertheless felt a personal piety. Swiss Protestant *philosophes* were notable for their own rational faith [Taylor, in 124]. Some scholars have suggested that even Gibbon may have been religious in his own, purely personal and private, fashion [119: ch. 5]. And most enlightened intellectuals believed that decency required a certain outward conformity to the public ceremonies of the established church, whether one believed in them or not. Such well-mannered behaviour helped to maintain proper social order and civil peace. Gibbon was probably an unbeliever, but he was also a dutiful churchgoer (following divine service with a Greek Testament, he said, improved his languages).

Many Enlightenment theorists in any case expected that a well-constituted society would possess a 'civic' religion, upon the model of ancient Rome – a faith designed to foster patriotism, community spirit and virtue. Gibbon remarked of that Roman religion, that the people believed it true, the philosophers thought it false, and the rulers knew it was useful. Islam made a better 'civic religion', he concluded, than priest-ridden and otherworldly Christianity [119: ch. 5]. Along similar lines, Voltaire was notoriously convinced that it was essential that one's servants – one's wife, too – should be pious, otherwise, lacking the fear of God, such people would steal the spoons or be unfaithful. Recognizing the utility of devotion, many *philosophes* not surprisingly advocated a two-tier religious system, with a simple, pure, rational religion for the elite, and a melodramatic faith to regulate the minds and hearts of the plebs. Such beliefs eventually found expression in the cult of the Supreme Being, a modern, rational, de-Christianized object of worship, concocted during the French Revolution.

Finally, we must remember that not a few *philosophes* were deeply religious. In his later years in particular, Jean-Jacques Rousseau waxed lyrical in praise of a piety which was emotional and spiritual rather than intellectual [68]. The great English polymath Joseph Priestley, 'discoverer' of oxygen, and uncoverer of the corruptions of Christianity, was second to none in his enlightened ardour that education, science, industry and technology would bring about endless progress; but he expressed his belief in human perfectibility in terms of a literally millennialist Christianity. Priestley passed most of his career as a Nonconformist preacher of Unitarian leanings. (Unitarianism, which denied not only the Trinity but the divinity of Jesus Christ, was, quipped Erasmus Darwin, a 'feather bed to catch a falling Christian'.)

Especially in the main Protestant regions – in Northern Germany, Scandinavia, England, Scotland, and the Calvinist cantons of Switzerland – advanced thinkers tended not so much to be hostile to Christianity *per se*, or to religion in general, but were rather concerned to achieve a purified, refined expression of faith, which would prove commensurable with reason and science, conscience and probability. Countless educated people could see no reason why such a faith ('true religion') should be an obstacle to progress. Just as the Lutheran Reformation had purged the Medieval church of its corruptions, so, they argued, the age of reason would complete the process, ridding worship and creed alike of those absurd Medieval accretions (such as angels and the literal belief in eternal hellfire) which progressive thinkers could no longer credit. Articles of faith that had been valuable props to devotion in darker days would naturally wither away with man's intellectual advance.

Pathbreaking in this respect was John Locke's *The Reasonableness of Christianity* (1695). Locke argued that the thinking man must be a believer, precisely because Christianity's central doctrines – belief in an omnipotent, omniscient, benevolent Creator, the duty of obeying and worshipping Him, and so forth – were all perfectly consonant with reason and experience. Being a Christian was a rational commitment; but the reasonable Christian was not obliged to accept features of traditional faith at which his reason baulked. No irrational leaps of faith were required. In the guise of 'rational religion', Christianity was thus being pared down to the minimum which educated people found easy to credit.

Rather more *avant garde* advocates of such forms of 'rational Christianity' further argued that the incarnation of Christ, with his evangelical mission, his healing miracles, etc., had not been strictly necessary for the imparting of essential religious truths to rational people. Such 'revealed religion', as recorded in the New Testament, had, however, been required (suggested Anthony Collins and other deistic 'freethinkers' who emerged in early eighteenth-century England) to sway the unthinking herd [128].

Of course, it was but a short step from this standpoint for radical freethinkers further to conclude that, when subjected to the acid tests of reason and history, Biblical Christianity failed all down the line. Were not the so-called 'miracles' of the Old and New Testaments mere fables and fictions, designed by crafty clergy to overawe the ignorant? Modern science would expose their trickery or explain their 'mysteries' away. Hence, freethinkers claimed, it was a waste of effort to try to reconcile Christianity with reason. Educated people should frankly admit that Christianity was inherently irrational, to be abandoned in favour of a rational alternative, commonly known as 'Deism'. Deists, such as Collins's Irish contemporary John Toland, and, for most of his career, Voltaire, contended that contemplation of the order of things led the mind, in Alexander Pope's phrase, 'thro' Nature up to Nature's God'; theirs was thus a purely 'natural religion' [9].

For some, of course, the step from Christianity to natural religion led further, to a 'religion of Nature' itself. This rested on the claim that there was no good reason to believe that any conscious, intelligent principle, any Supreme Being or Great Architect, lay behind and beyond Nature at all. Nature was all there was, and in so far as anything was sacred, and requiring worship, it was Nature herself. The great seventeenth-century philosopher Spinoza had influentially suggested that 'God' was the same as 'Nature'. A similar sort of atheism, or materialistic pantheism, was later boldly expressed in the Baron d'Holbach's *Système de la nature* (1770) [84; 88].

Eighteenth-century intellectuals, in other words, tried to find or forge a religion fit for the times. Traditional Christianity was widely found wanting, at least by and for the educated. Some tried to rationalize and refine it, others to create a more credible alternative.

The question remains, however, why so many *philosophes*, the French in particular, expressed such vitriolic hatred towards the Christian religion and church, habitually satirizing priests as perverts, friars as gluttons, monks and nuns as lechers, theologians as hair-splitters, inquisitors as sadistic torturers, and Popes as megalomaniacs.

In part it was because they convinced themselves that organized Christianity was a cold, calculating fraud. Churchmen, they often hinted, did not even themselves believe all their mumbo-jumbo, but, like cunning conjurors, knew very well that long words in Latin, sleight of hand, and pomp and circumstance conferred power over the people. Explaining the rise of the Church under the Roman Empire, Gibbon pictured the early Christians as a gang of ruthless zealots, intent upon aggrandizing themselves, no matter what cost to social peace. By cynically cooking-up doctrines such as purgatory and the system of pardons to go with it, the Medieval church had manipulated minds, waged war upon its enemies, and become Europe's richest multinational organization.

The Church, as outraged *philosophes* saw it, had thus been not merely mistaken or unscrupulous, but positively evil. Hypocritically preaching peace, it had sown discord and strife. The religious wars of the sixteenth and seventeenth centuries had spilt oceans of blood. Every year the anniversary of the St Bartholomew's Eve massacre made Voltaire physically sick.

Even in the eighteenth century, so the *philosophes* believed, the perverted teachings of Christianity were still warping minds, for instance by seducing young men and women into joining monasteries and nunneries, and by gratuitously terrifying children with fears of damnation. Here the Calas case (1762) offered perfect publicity to the *philosophes*. The Calas family were Toulouse Protestants. Their eldest son was found dead. Rumour had it that he had been planning to convert to Catholicism, and that, to prevent this 'disgrace', his father had murdered him. A trial of dubious legality found the father guilty, and he was executed.

Voltaire took up the cause. For him, the affair revealed the monstrous evils of religious bigotry, whatever precisely had happened. If the father had indeed murdered his son in the name of the upholding of the Protestant faith, it proved how grotesquely religious sectarianism could undermine family feeling. If the father were innocent – as Voltaire, of course, believed – it showed the malice of confessional strife amongst those who prided themselves upon being the bringers of peace. Cases such as this enabled *philosophes* to quote the Roman poet Lucretius: '*tantum religio potuit suadere malorum*' (how great the evil which religion induces men to commit).

What angered the *philosophes* above all was that churches – opulent, and a drain on the economy – were still exercising mind-control and political power. The Catholic Church in particular continued to outlaw other faiths. It largely monopolized the education system, from infant schools to seminaries and universities (both Voltaire and Diderot had had excellent Jesuit educations, and never forgot it). It censored books: most of the *philosophes'* works found their way onto the Index of Prohibited Books. In such *causes célèbres* as the trial of Galileo, the Church had arrested the progress of knowledge. Some one hundred and fifty years later, in the 1770s, the leading French natural historian, Buffon, was still being required to answer to the holy fathers of the Sorbonne for arguing that the Earth was much older than the Bible implied. Where Throne granted Altar the sword, as in Spain, the result was appalling intellectual and scientific stagnation. Only where the wings of churches had been clipped by the civil authorities, as notably in the Dutch Republic and England, was progress assured.

Ultimately, then, the ferocity of the *philosophes'* onslaught arose from their own personal experience and circumstances. From the 'dark ages' onwards, they argued, the clergy had dictated the intellectual life of Europe and exercised a mind police. But things were changing. With the rise of literacy, the spread of education and the greater circulation of books, a new secular intelligentsia was flexing its muscles and finding its feet, challenging the clergy for the ear of the people. The *philosophes* saw themselves as the advanced guard of this body of writers and thinkers. They were demanding free expression for themselves. They aimed to replace the clergy as the mouthpieces of modernity.

In its campaign against *'l'infâme'*, the Enlightenment offered a new deal for the European mind. *Philosophes* demanded an end to censorship, and celebrated the printing press as a genuinely liberating technology. Yet in their turn, they also ironically mirrored the clergy they were aiming to supplant. They too formed their cliques, their 'holy circles'; often they too cultivated a taste for secrecy, and some took great pleasure in developing their own intimate rituals and occult symbols. Not least, many *philosophes* were early and enthusiastic members of freemasons' lodges, which were newly emerging at this time. Such lodges were secret gentlemen's clubs, congenial centres of fraternal solidarity, which bound their members with bizarre entrance rituals that sometimes blasphemously parodied the rites of the church [84; 85]. Some *philosophes* believed in the need for mass religion as an 'opium of the people'. And the very slogan *'Écrasez l'infâme'* itself echoed the bloody war-cry of the crusader, only this time that of the '*philosophe militant*'.

In some ways, therefore, the 'little flock' of the *philosophes* could be said to be creating a new religion of its own, a religion of humanity. It remains true, nevertheless – as will be further discussed in the final chapter – that the eighteenth century marks a major stage in the secularization of Europe, a development for which the *philosophes* at least provided all the main arguments.

5 Who was the Enlightenment?

There was a time when historians commonly explained views by reference to the 'spirit of the age', the *Zeitgeist* or *Weltanschauung* – a notion perhaps ultimately derived from Hegel's idea of the successive dialectical stages of the fulfilment of consciousness in history. Features of Enlightenment thinking were thus said to have been expressions of the 'rationalist' or 'individualist' temper of their times. Explanations of this kind nowadays appear tautologous.

Hardly more satisfactory to us today is the recourse Marxist historians traditionally had to such slogans as 'bourgeois ideology', when attempting to explain how it was that many Enlightenment thinkers believed that society was no more than the sum of its component parts, or who thought of morality, in the manner of Jeremy Bentham, as nothing more than a quantifiable calculus of pleasures and pains, translatable into cash values [16; 60; 51]. Marxists classically identified the eighteenth century as marking the final transition from feudalism to capitalism, culminating in bourgeois revolution (the French). The Enlightenment was thus understood as the manifesto of the 'bourgeois' struggle against 'aristocratic' feudalism [63].

The problem with such formulae is that they explain both too much and too little. They will hardly tell us why a man of impeccably middle-class origins such as Rousseau, a watchmaker's son, coming from that ultimately bourgeois city, Geneva, repudiated the 'bourgeois ideology' of liberal individualism contemporaneously being adumbrated by such *philosophes* as the Marquis de Mirabeau and the Baron d'Holbach, whose titles placed them squarely within the ranks of the nobility [27]. Neither will they explain (as will be discussed below) why the French mercantile community seems to have been notably indifferent to the Enlightenment's supposedly 'bourgeois ideology'.

Dissatisfied with such pseudo-solutions which promise much and deliver little, many of the leading historians of ideas and philosophy in the first half of the twentieth century abandoned sweeping generalizations about class and class-consciousness, and instead set about exploring in depth the internal structures of theories and their relations to each other. They were primarily interested, not in who the *philosophes* were, or what they did, but in the ideas they formulated, and how they influenced each other. More specifically still, their primary concern was not the cut-and-thrust of polemics and rhetoric, but the underlying coherence and interconnections of their thought-worlds, and the (sometimes unstated) metaphysical and philosophical postulates upon which they rested. In his *The Philosophy of the Enlightenment*, Ernst Cassirer, doyen of this approach, demonstrated how the late seventeenth-century empiricist revolt against traditional philosophical rationalism – led by Locke, and taken over by Condillac, Helvétius and others – in turn descended into a baffling scepticism (expressed above all by Hume), which it took all Kant's philosophical acumen to resolve [35; 28; 59].

The problem with histories such as Cassirer's is that they blithely disembody ideas and evaluate them by timeless criteria, outside their wider historical context. Living people are reduced to doctrines on the printed page, in ways the more materialist of the *philosophes* themselves ridiculed. They also tend to impose their own retrospective and often ahistorical evaluations as to who was 'truly' important (which usually means those making contributions of permanent value to the solution of philosophical problems). In this respect, Cassirer's book is notable for not even mentioning such crucially influential figures as Adam Smith and Jeremy Bentham, since they were evidently not sufficiently philosophically profound [35; 1]. The approaches associated with the French historian of systems of thought, Michel Foucault, and later with Postmodernism, do something similar. Denying the conventional idea of the 'author' – on the grounds that we can never get into the head of a writer – these emphasize that history cannot go beyond the analysis of 'discourse' or texts ('textuality') [55; 126].

Peter Gay's *The Enlightenment: An Interpretation*, by contrast, was far more concerned to bring the *philosophes* alive, as practical pundits rather than doctrinal mouthpieces and carriers of ideas – people whose opinions and polemics arose out of living the lives of embattled intellectuals, with particular personalities and psychological dispositions. Gay advocated what he called the 'social history of ideas' [60]. Yet, as we have seen, he still chose to focus his attention on what he termed a 'little flock', a 'party of humanity' – above all, Voltaire, Diderot and Rousseau. As a result, the French, or French-speaking, Enlightenment was granted great, and perhaps undue, prominence (a perspective decentralizing France will be offered below in Chapter 6). His book perpetuated, almost despite itself, the 'great men, great minds, great books' bias [61].

More recently, historians have attempted to advance Gay's programme of a 'social history of ideas' beyond Gay's own horizons. They have focused upon the fluid and fruitful interplay between wider groupings of thinkers, and their ebbings and flowings; they have emphasized how the Enlightenment was a broad collective endeavour, not just the work of a few giants: and they have underlined the intimate interconnections between material circumstances, people's lifestyles, and their thinking. This newer approach incorporates the concern with the material basis of consciousness which Marxism rightly pioneered, but abandons rigid formulae (for instance, the Enlightenment seen as the struggle of emergent bourgeois society to break free of its feudal shackles) in favour of local knowledge of activists, groupings, and crises [47; 84; 109].

Most fruitfully, this new 'social history' of the Enlightenment has contended that it is a mistake to date the Enlightenment from the emergence of its 'big books' of the 1720s and 1730s: Montesquieu's *Lettres persanes* (1721), Vico's *Scienza Nuova* (1725), Voltaire's *Lettres philosophiques* (1733). Neither may we take it for granted that it came to an end with the deaths of its heroes: all the great French *philosophes*, except Condorcet, died before the French Revolution.

Instead, recent social historians have invited us to regard the movement as a wider ferment inaugurated, sustained and spread by a vastly larger number of relatively obscure thinkers, writers, readers and contact loops. Nor could it ever have flourished without extensive support-networks of friends, sympathizers and fellow-travellers – comrades who gave refuge to exiles, or passed on letters and books to those living underground, in hiding. Sometimes the Enlightenment even drew upon the connivance and clandestine aid of those in authority, prepared to turn a blind eye to what were illegal publishing activities. Thus through his liberal policies, Malesherbes, the head French censor at mid-century, perhaps proved the French Enlightenment's best friend. Above all, the Enlightenment would probably have fizzled out without the intrepid, and often highly risky, support of printers, publishers and book-distributors, who often had to organize the smuggling of illegal books across borders. Leaders of the book-trade were often willing to run great risks to publish the works of the *philosophes* – occasionally out of political conviction, though generally, Darnton has emphasized, with tidy profits in mind [46; 47; 87].

'Underground' activities of this kind were crucial to the earliest groupings of Enlightenment activists, and led to the expression of vicious satire and extreme views, almost more radical than anything that came after. As Margaret Jacob has emphasized, the fierce political and religious criticisms characteristic of the early Enlightenment arose largely out of resistance to what appeared to be the rampant alliance of rising Bourbon and Stuart absolutism, yoked with Catholicism, which threatened to carry all before it in the last quarter of the seventeenth century [84].

Politico-religious exiles from the England of Charles II and James II, and Huguenot refugees from France (particularly after the Revocation of the Edict of Nantes in 1685), joined up in the Dutch Republic with native radicals, freethinkers, men of letters, publishers and printers. Taking shelter in Utrecht, Rotterdam or Amsterdam in an atmosphere of deep uncertainty and intrigue, such men set up their own informal, clandestine organizations, clubs and circles, which sometimes gravitated in the direction of freemasons' lodges. Such coteries celebrated the 'republican' virtues of brotherhood and freedom, for their exiled members were compelled to make 'republican' comradeship a way of life. These little republics of letters excoriated political absolutism, developed their own critiques of orthodox religion, based, above all in the writings of John Toland, upon a pantheist creed in which Nature replaced the Christian God, and in which (in

41

certain manuscripts circulating clandestinely) Moses, Jesus and Mohammed were equally exposed as 'impostors'. The writings of the exiled Toland – republican in politics, materialist in philosophy – form the most coherent expression of the more extreme views arising from this milieu [84; 155; 156].

Precisely how to evaluate this early 'radical Enlightenment' and its wider influence remains highly contested amongst scholars, partly because the surviving evidence is so fragmentary and difficult to evaluate. What is clear, however, is that some of the most extreme repudiations of politico-religious orthodoxies had already been formulated at the very outset of the Enlightenment – views which make early Voltaire look tame by comparison. As Ira Wade showed long ago, such ideas quickly spread, through the clandestine circulation of manuscripts, above all, into France, to fertilize later radicalism [155; 156].

It is becoming clearer that, at all stages of the movement, nine-tenths of the Enlightenment iceberg was submerged, and that the historian neglects this at his peril. Scholars have long lavished great attention upon Diderot and d'Alembert's *Encyclopédie*, that great compendium of practical knowledge spiced with daringly advanced views, which began to appear in 1751, and finally ran, over the course of some twenty years, to some twenty volumes, with a further ten volumes of plates. The authors of certain of the contributions remain unknown to this day; many – not least the Chevalier de Jaucourt, who wrote hundreds of entries – receive hardly a mention in standard Enlightenment histories. All stand witness to the indispensable reservoir of supporters and sympathizers upon whose services top *philosophes* could rely. Much of the *Encyclopédie* – that 'Trojan horse' of the *ancien régime* – comprised accounts of arts and crafts, science and technology, industry and agriculture, all of which the editors believed were vital for the modernization of France. Yet a sprinkling of entries was much more openly subversive, and d'Alembert's own 'Preliminary Discourse' proclaimed in ringing tones the need for new ways of thinking to meet the needs of a new age [92; 125].

Elaborate study of the contents of the *Encyclopédie* would be meaningless, however, without a grasp of its history as a book: how it was financed, published and circulated, who distributed it, who bought it, read it, and with what effect. Robert Darnton has immeasurably improved our understanding of these issues with his in-depth study of the numerous re-publications of the *Encyclopédie* in the latter part of the century, usually in smaller, cheaper formats. In this enterprising activity, one particular Swiss firm, the Société Typographie de Neuchâtel, was specially prominent. It specialized in putting out pamphlets and *risqué* fiction written by French authors, which it was still hazardous to print in France. It is, of course, highly significant that publishers in Paris, too close to the watchful eye of authority, and accustomed to enjoying their own privileges, were in no position to reap the big profits from popularizing the works of the *philosophes* [45; 47].

So who bought the *Encyclopédie* and similar works? Certainly not the 'people' or even droves of lower-class radicals: it was too expensive, a radical work aimed at the rich [37]. Nor – despite the attention given to trades and technology – were its buyers the commercial bourgeoisie. French businessmen were too interested in time-honoured ways of making money and gaining status, and were to prove amongst the more conservative elements of the late *ancien régime* [27]. Rather, the main market came from the upper professional classes (lawyers, administrators, and office holders), the higher clergy, aristocratic landowners, and provincial dignitaries. It was these affluent, influential, and educated circles who also made up the membership of the literary academies and learned societies which became so prominent in the French provinces in the second half of the century [130; for an English comparison, the Lunar Society of Birmingham, see 140]. Paradoxically, it was upon the patronage and purses of these pillars of the establishment – people at bottom socio-politically quite conservative, though often with an eager appetite for intellectual novelty and fashionable culture – that the Enlightenment itself was sustained.

The high noon of the 'High Enlightenment' in France thus involved an elite of intellectuals, chiefly in comfortable financial circumstances, writing for members of their own class. Montesquieu was a baron, Condorcet a marquis; Voltaire, a lawyer's son, grew extremely rich upon the profits of writing, and lived in style as the *propriétaire* of the chateâu of Ferney. Helvétius made a fortune as a tax-farmer. Gibbon, son of a landowning MP, succeeded his father in Parliament. Bentham lived off inherited wealth. For all their praise of manual skills, few *philosophes* were truly self-made men, sprung from the people – though Diderot was, as was the Philadelphia printer Benjamin Franklin, who followed the advice he published in his *Poor Robin* almanacs ('Early to bed, early to rise, makes a man healthy, wealthy and wise') [17]. The most outrageous freethinker of all the *philosophes*, the Baron d'Holbach, was a German-born aristocrat, who kept a lavish Paris salon [87].

It was only in the latter years of the century that fertile interplay – and, eventually, electric tensions – arose between Enlightenment intellectuals and popular culture in France. As Darnton has shown, a new sort of Enlightenment made its appearance, especially in the 1780s: an embittered style of popular journalism. Its hack-writers did not merely savage in crude and rabble-rousing terms such traditional sitting targets as priests, tax-farmers and court-iers, but also showed visceral hatred towards the privileges of high society at large. This new Grub Street journalism pioneered in France the techniques of the gutter press, going in for muck-raking and the exposure of sexual scandal, but also peddling simplified Enlightenment slogans in the process. At a time when the *philosophes* had largely been absorbed into fashionable society, new writers – for instance, Sébastian Mercier and Restif de la Bretonne – took the politics of Enlightenment to the disaffected and propertyless, beating out more discordant messages, which looked, for their patron saint, not to the ultra-sophisticated Voltaire, but, significantly, to that populist railer against polite society, Rousseau [47; 54].

Peter Gay has characterized the *philosophes* as a 'solid, respect-able class of revolutionaries' [61: vol. 1, 9]. This deliberately provocative formulation has a broad ring of truth to it. However radical their ideas, the leaders of the 'High Enlightenment' came from, or rose into, the polite and propertied classes and thought of themselves as genteel. In England, the members of the Lunar Society of Birmingham, that lustrous Enlightenment gathering of intellectuals, doctors, entrepreneurs and inventors – it included the manufacturers Josiah Wedgwood and Matthew Boulton, the inventor James Watt, and the scientists Erasmus Darwin and Joseph Priestley – celebrated progress, deplored slavery, and saluted the outbreak of the French Revolution. Yet none took up the cause of the 'people'; indeed, it was a popular church-and-king mob which burned down Priestley's house in 1790, and drove him into exile in the United States [96; 139].

I have been using the phrase 'men of the Enlightenment'. But what of its women? It is easy to find females who played a certain role in the movement: the Marquise de Châtelet, Voltaire's companion, who had a fine grasp of Newtonian science, or Sophie Volland, Diderot's mistress – highly intelligent, cultured and articulate. Belle de Zuylen, who became Mme de Charrière and settled in Switzerland, was a talented literary lady with advanced views – James Boswell, who enjoyed a brief flirtation with her, concluded she was a 'frantic libertine', because she rejected the sexual double standard. Such 'blue stockings' as Elizabeth Montagu and Mrs Chapone held court in London [106], and, as Gibbon found, *grandes dames* ran the salons in Paris at which the *philosophes* shocked and scintillated. Dena Goodman has contended that such salons formed the key site for the exchange of enlightened ideas [64; for doubts see 101]. The massive expansion of print culture gave women new opportunities as both readers and writers [50; 82; 146; 148].

Yet Gay's 'family' of *philosophes* is a men-only affair. It can be argued that no enlightened woman emerged as a front-rank innovative philosopher, scientist or intellectual until the Revolutionary period, when Mary Wollstonecraft (author of *A Vindication of the Rights of Woman*, 1792) and, slightly later, Madame de Staël made their mark [147; 149]. The Church had permitted women to express themselves, as saints and mystics, and, of course, many royal and noble women exercised great power in the *ancien régime*, Maria Theresa and Catherine the Great being leading examples. But it was not till the nineteenth century that we should speak of authentic 'women's movements', led by women for the advancement of the sex [54].

The (male) Enlightenment warmly and broadly encouraged the view that women ought to be treated as rational creatures, and such authors as Locke held that girls should have more or less the same education as boys (that followed from the notion that the mind began as a *tabula rasa* – the mind had no sex [138; 41]. But beyond that, *philosophes* did not generally commit themselves to the general emancipation of women as men's equals. While they complained against prejudice and injustice, hardly any women thought in terms of enfranchisement and political participation, or the opening of the professions to their sex. Indeed, advanced female thinkers, like Mary Wollstonecraft, especially praised

45

women's role as mothers and educators of children: it was for that reason that women deserved the best of educations and the highest social respect.

Certain *philosophes*, notably Rousseau, espoused a clear-cut sexual division of labour [138]. Public life should be for men, while women (regarded by the Genevan as creatures of feeling) were to tread the honoured but obscure path of private virtue, modesty, domesticity and child-rearing. Such views – with their fears of 'effeminacy' and of 'bedroom politics' – were widely reiterated in the rhetoric of the French revolutionaries, who had the greatest trepidations about women meddling in public affairs, and they became embedded in the nineteenth-century notion of 'separate spheres'. In valuing reason, yet also helping to launch a cult of idealized motherhood, the Enlightenment left an ambiguous legacy for women [41; 78; 82; 147].

6 Unity or Diversity?

Amongst the values dearest to Enlightenment writers was cosmopolitanism. Claiming that reason, like the Sun, shed the same light all the world over, the *philosophes* commonly insisted that there was a single universal standard of justice, governed by one normative natural law – and indeed that there was a single uniform human nature, all people being endowed with fundamentally the same attributes and desires, 'from China to Peru'. Thus one of the favourite literary genres of such writers as Montesquieu was to assume the persona of a foreign 'anthropologist' (for instance, a Persian sage) visiting Europe – as a way of satirizing the vices and follies not just of Europe but of mankind at large [139].

The *philosophes* mocked narrow-minded nationalism along with all other kinds of parochial prejudice. They liked to view themselves as men of the world, who belonged less to Savoy, Switzerland, Scotland or Sweden than to an international republic of letters. In so doing they looked back admiringly to the internationalism of the Stoic philosophers of Antiquity, and to the common civilization which had united the Mediterranean in the days of the Roman Empire. Science and scholarship, after all, still drew upon a common learned language, Latin, for their communications. And if Latin was losing ground in the eighteenth century, a new international *lingua franca* was taking its place – French. Frederick the Great communicated with Voltaire in French as did Catherine the Great with Diderot; Edward Gibbon even published his first book in that language.

Here lay certain ambiguity. The French tongue came to serve as a kind of Esperanto. But its dominance also marked the fact that, for many, French intellectuals were in the vanguard of advanced thought. And historians, while asserting the internationalism of the Enlightenment, have often at the same time assumed that it was developments in France which were definitive, and that light shone out from there to illuminate all the rest of Europe. For instance, Leonard Marsak has claimed, along these lines, that the Enlightenment 'was primarily a French phenomenon' [11; 6], while Robert Darnton has recently re-stated that it was 'in Paris in the early eighteenth century' that Enlightenment took off [49]. The further away from Paris, the darker things got.

But there are dangers in making too close an identification of 'true' Enlightenment with intellectual and cultural activity in France. For one thing, as we saw in the previous chapter, the initial centre of the Enlightenment ferment was certainly not France. Before the emergence of towering geniuses such as Montesquieu and Voltaire, a radical intellectual movement had bubbled up beyond the French borders, especially in association with the seventeenth-century Dutch Republic and with post-1688 England. Hence we need to look to the origins of the French Enlightenment.

For another, throughout Europe, the leading intellectuals, while of course subscribing to cosmopolitanism, were deeply engaged in getting to grips with local problems and politics; they had their own distinctive priorities for handling them. Hence, we will end up with a tremendously distorted picture of the Enlightenment if we assume that the issues preoccupying French *philosophes* – the evils of Catholicism, 'feudal' privilege, censorship, the need to develop philosophical materialism – automatically counted most in the minds of thinkers the length and breadth of Europe, from Naples to Uppsala, from Birmingham to St Petersburg. Instead, we will find that intellectuals addressed themselves to the problems of their own societies and regions, developing 'enlightened' solutions from within their own cultural values.

Dutch thinkers resolved the problems of the Enlightenment almost before anyone had experience of them. As Schama has emphasized, the seventeenth-century Dutch Republic was *sui generis*. First the Spanish, and then the French, tried to blast the nation off the map by force of arms. They failed, but even so traditional wisdom held that the United Provinces was such a peculiar entity that it had no business to survive. For it harboured a multitude of different peoples and creeds – not just Protestants and Catholics, but Jews and heretics as well. It was, furthermore, a political hybrid, possessing, in the *stadholder*, only the palest imitation of a monarch, and relying upon a devolved, ramshackle, republican political structure which was often divided among itself. It was dominated less by the conventional hereditary military aristocracy than by burghers, and its wealth came not from the land but from trade. Nor was it a 'republic' in the classical mould. For political wisdom about republics demanded that they should cultivate frugality and renounce 'luxury' [25]; the citizens of Holland, by contrast, enjoyed their uncommon prosperity to the full [Schama, in 124; 137].

The Dutch Republic was thus, in the eyes of many, a nonsense, and no-one knew quite what to make of it. Yet against all the odds, by the close of the seventeenth century it had clearly proved a success story, and thereby became a shining instance of Enlightenment desiderata in operation: freedom from tyranny, religious pluralism and tolerance, prosperity, a (relatively) peaceful foreign policy. Its safe-haven for intellectual exiles, its scientific productivity, and its high-quality publishing houses made the Dutch Republic seem like an enviable recipe for progress.

And yet the tributes to the dazzling seventeenth-century Dutch experience remained muted and grudging. No *philosophes* seriously thought that France, Prussia or Russia could or should follow the Netherlandish political model. Aristocratic *philosophes* despised the Dutch – as they commonly despised the Jews – as dirty, money-grubbing tradesmen. They were not surprised that as the United Provinces grew more oligarchic in the eighteenth century, dominated by rich bankers and international merchants, its cultural achievements faded away. By 1750, the Dutch had ceased to make an independent contribution of their own to science or philosophy, remaining only a key centre of the publishing trade [Schama, in 124].

The uniqueness of the British* experience could not so readily be dismissed. Admittedly, eighteenth-century England did not produce that galaxy of daring intellectuals, radiating all that was most radical in politics, freethinking, and moral and sexual speculations, which flourished in France [124]. Yet this was not because Georgian England was benighted. Far from it. It was because England was already undergoing, before the eighteenth century

---

*The Act of Union of 1707 dissolved the Scottish Parliament and united the thrones of England and Scotland.

opened, those transformations in politics, religion, and personal freedom for which French and other radicals had to clamour, unsuccessfully, all the century [116; 117; 118].

Above all, because of the 'Glorious Revolution' of 1688, England had achieved the guarantee of parliamentary representation and constitutional government, individual liberty (*habeas corpus*), substantial (though not complete) religious toleration, and freedom of expression and publishing. John Locke and his followers had in effect produced blueprints for the enlightened society: a liberal regime based upon individual rights and natural law, the priority of society over government; a rational Christianity; the sanctity of property, to be deployed by owners within a liberal economic policy; a faith in education; and, not least, a bold empiricist attitude towards the advancement of knowledge, which championed the human capacity to progress through experience [165; 99].

The grand problem facing English intellectuals in the Georgian century lay not in the need to criticize an old regime, or to design a new one at the drawing-board, but rather in defending their reformed polity and making it work. It was a bold experiment. Could a large measure of individual liberty prove compatible with socio-political stability? Or would limited constitutional government collapse into either anarchy or despotism? To prevent these, the virtues of the checking and balancing mechanisms of a mixed constitution were widely extolled – devices soon to commend themselves to Montesquieu and to be incorporated into the American Constitution. It was crucial, however (so Hume insisted) that the excesses of party political rant and rhetoric should be tempered by the wiser moderating counsels of experience [Porter, in 124].

Likewise – no less of a problem – could the vast surge of individual prosperity in the age of imperial expansion and industrialization prove compatible with social cohesion? Or would wealth subvert liberty, divide class against class, and corrupt the constitution – all dangers forefronted in traditional 'commonwealth' ideology [116]. Once again, an optimistic alternative was formulated, from the witty paradoxes of Bernard Mandeville early in the century to the systems of Adam Smith and other political economists towards its close [81]. This contended that the wealth of individuals would successfully enhance the wealth of nations, and that prosperity inevitably wove webs of interpersonal connections which strengthened, rather than divided, society [71].

Moralists nevertheless feared that what has been called 'possessive individualism' (the pursuit of private gain), let loose in an 'opportunity state', would prove too disruptive, leading to the alienation of man from man [97; 116]. On the contrary, counter-argued a major current of British thinking, from Addison and Steele in the *Spectator* early in the century, through to such late Enlightenment Scottish professors as Adam Smith, John Millar and Dugald Stewart [94; 38]. Economic progress would produce a consumer society which would, in turn, serve to refine manners, promote peace, soften sensibilities, and bind men to their fellows by the invisible chains of commerce. Properly understood, commercial capitalism and its attendant urban living would prove not the solvent of society but its very cement. Leading British intellectuals were thus more preoccupied with practicalities than with abstract programmes [114; 115; 31; 24; 25].

This applies above all to the extremely distinguished body of men of letters and professors who made up the Scottish Enlightenment. Early eighteenth-century Scotland was economically backward. The Act of Union (1707) dissolved the independent Scottish Parliament. The failure of Jacobitism further divided and weakened the nation. Such great Scottish thinkers as David Hume and Adam Smith responded not by wringing their hands, basking in former glories, or engaging in visionary utopian schemes for independence. They recognized rather that Scotland's future would depend upon rapid social modernization and commercial development. Their pioneering analyses of the social preconditions of capitalism, and of the laws of a capitalist economy, constitute the great achievement of the practical genius of the Scottish Enlightenment [31; 38; 82].

If Continental intellectuals had mixed feelings about the Dutch marvel, they were hardly in doubt about the British. 'Anglomania' swept the continent, fired by Voltaire's *Lettres philosophiques* (1733), which positively glowed about Britain's political liberty, religious toleration, economic success, cultural modernity and scientific glories – Newton above all. English cultural innovations, notably periodicals, like the *Spectator*, and novels, from Defoe's best-selling *Robinson Crusoe* onwards, were widely imitated. In exile in England, Voltaire had seen the future and it worked.

51

The English state was a low-key constitutional polity, in which the Crown operated in a complex permanent partnership with the two Houses of Parliament. Many believed this limitation on central power was an admirable recipe for encouraging civil society (merchants, craftsmen, artisans) to blossom into self-sustaining economic and cultural growth. Yet as a system it was hardly exportable.

After all, in most of the German principalities and the monarchies east of the Elbe, a feudal aristocracy remained in the saddle, the rural economy was backward and torpid, literacy levels were abysmal, and enterprising technocrats and industrialists, capable of generating economic change from within, were few. Hence, in most parts of continental Europe, an alternative model for spreading Enlightenment was needed – one which, instead of (as in England) relying upon market mechanisms and individual initiatives to generate improvement, and instead of assailing the monolithic *status quo*, as in France, attempted to conscript existing institutions and to reform them in the process.

In Southern Germany and the Habsburg lands, the Roman Catholic Church undertook its own reformist programme, promoting a more conciliatory faith, divorced from the dogmatics of the Counter-Reformation [Blanning and Wangermann, in 124]. In Protestant Germany and Scandinavia – as also in Switzerland and Scotland – the universities served as the agents of intellectual ferment; professors enunciated a rational Christianity, produced schemes for economic regeneration and administrative reform, and eagerly took up the cause of science [Frängsmyr, Phillipson, Taylor and Whaley, all in 124]. Many were involved in official surveys of national natural resources. If Gibbon thought Oxford dons were sunk in port ('their dull and deep potations excused the brisk intemperance of youth'), and Voltaire lambasted the Sorbonne as closed to inquiry, the universities at Uppsala, Halle, Göttingen and elsewhere offered a foretaste of the astonishing recovery of academic life that was to take place in the nineteenth century.

Not least, royal circles themselves became committed to enlightened government. The so-called 'enlightened absolutists' – Frederick the Great, Catherine the Great, possibly Maria Theresa and certainly her son, Joseph II – recognized the need for reform, if only to improve their fiscal base and modernize their fighting machines [89]. They promoted education, updated taxes, encouraged trade by abolishing internal tolls, and built up bureaucratic hierarchies designed to perfect rational, efficient and orderly management. *Cameralwissenschaft*, the 'science' of administration expounded by such jurists as Justi and implemented by such great state servants as Count Kaunitz, was the practical political science of the Continental absolutisms, designed to produce the 'well-ordered police state' ('police' here having the connotation of 'bureaucratic', not its modern analogue, 'fascist') [127].

In central and eastern Europe, the educated, professional and official classes (the *Beamtenstand*) fell in with the progress of bureaucratic Enlightenment from 'above'. Progressive noblemen experimented on their estates with agricultural modernization, and, as in Bohemia, founded societies for spreading science and technical know-how [Teich, in 124; 23]. Genovese and other Italian intellectuals instructed their compatriots in economic modernization [Chadwick, in 124; 152]. In cities from Copenhagen to Milan, newly-founded reading societies gathered over the new beverage, coffee, to hear essays on history, philosophy and future prospects, and to acquire the virtues of polite culture, useful learning, and refined taste in the arts, which Addison and Steele's much-admired and imitated *Spectator* magazine had done so much to promote early in the century [94].

Enlightened activity in central Europe operated within existing political structures. Unlike in France, its priority was not to assail church and state and call for radical political liberties. Such strategies would have seemed shallow, foolish, and, above all, foredoomed. But if a certain political conservatism remained the order of the day, we should not belittle this ferment of ideas. Everywhere, thanks to the springing up of new journals, newspapers, libraries and clubs, the educated classes began to criticize old modes of living, and open the prospects for future agitation [Whaley, in 124].

In Britain, advocates of Enlightenment largely accepted the post-1688 settlement, defended it, and got on with the business of improving society through piecemeal economic change and personal betterment. In the British Colonies of North America, Enlightenment became the idiom of rebellion, and the founding fathers had a unique oportunity to set up an enlightened polity from scratch [Pole, in 124; 103; 111]. In central Europe, the

Aufklärer threw in their lot with the authorities, seeing in them the agents of orderly well-administered progress. It was in France, however, that the configurations of Enlightenment proved most unstable [Hampson, in 124].

Far more than in England, French *philosophes* deplored the way their kingdom was constituted: it was backward and repressive, a failure even upon the international stage. Nor did they feel loyalty towards it, unlike many *Aufklärer* towards their own principalities. With such rare exceptions as Turgot, the French regime did not reward them with political office and power.

But we must not be tempted, as a result, to regard the leading *philosophes* as hopelessly alienated intellectuals, penniless bohemians starving in garrets, or outlawed conspirators forced underground, with nothing to lose. Far from it. Many of them, conspicuously Voltaire, grew rich, famous and fêted as the unacknowledged, alternative legislators of their land, a kind of government in quasi-exile. If Louis XV and Louis XVI fitfully listened to, fitfully challenged, but largely ignored such would-be enlightened politicians as the Physiocrats (those pioneers of market economics), fashionable, educated and polite society (*'les gens de culture'*) by contrast bought their books and elevated their reputations [130; 46; 47]. Therein, of course, lay the *philosophes'* strength – their fame protected them from being silenced – but also their weakness.

Distinctive relations grew up everywhere, between church and nobles (the first two Estates of the realm), the people (the Third Estate), and the advocates of Enlightenment (articulate opinion, the Fourth Estate). No fatal socio-political tensions were produced by the Enlightenment in central and eastern Europe, or in such small absolutist realms as the Italian duchies or Portugal. There the tendency was for the intelligentsia to be incorporated within the hegemonic order.

In England, by contrast, such tensions as existed did not reach breaking-point, because the state had already conceded liberty of expression and plenty of scope for the development of civil society and the economy. The activities of independent writers, propagandists, critics, industrialists and so forth were no real threat to the state. English intellectuals and artists, while often vocally anti-king or anti-ministry, profoundly identified themselves with the cause of the nation at large, and tended to be effusively patriotic.

54

France was the great anomaly. It was a society sufficiently modern, literate and wealthy to possess an influential intelligentsia, confident of its own muscle-power, and relatively independent of Crown patronage, ecclesiastical preferment or academic honours. French thinkers were not afraid of being crushed by authority. Yet it was an intelligentsia disaffected and capable of speaking to the disaffected. It would be extravagant to imply that the French Enlightenment brought about the French Revolution. But the movement certainly helped to create a situation in which ideological loyalty to the old regime was eroded and the regime destabilized.

7 Movement or *Mentalité*?

In 1768 Louis Bougainville set foot upon the Pacific island of Tahiti, newly discovered by Europeans. As a cultured man, the French naval commander quickly published an account which praised the island for its similarities to the Isles of the Blessed as evoked by the writers of Antiquity, a haven of ease, peace and plenty, where Nature spontaneously met all of man's wants. Above all, he stressed that these happy islanders held no private property, and that they were free of the rigid sexual taboos upheld in Christian Europe.

On reading this account of this paradise, the French *philosophe* Denis Diderot turned armchair anthropologist, penning a *Supplement* to his *Voyage*. Diderot went beyond Bougainville, in imagining Tahitian society as essentially free from all the curses of despotism and private property ('no king, no magistrate, no priest, no laws, no "mine" and "thine"'). All was held in common, including women, for this happy island, wrote Diderot, celebrated free love. The outcome was far from the degrading sensuality that Christian preachers would have predicted. The absence of fearsome prohibitions upon natural desires, rather, produced a society tranquil, gentle and psychologically well-balanced. Diderot praised these 'noble savages', while denouncing the warped and negative sexual attitudes of supposedly civilized Europe [58; 120; 129; 136; 164].

On his own second voyage to Tahiti, Captain Cook took issue with what he reckoned to be the glamourized and falsely sentimental picture of Polynesian life given by the French. It was an insult to the Tahitians, he contended, to portray them as wallowing in a form of communism: they were not so primitive at all. A careful observer, not dazzled by these fanciful preconceptions, would note (recorded the down-to-earth Yorkshireman) that almost every tree upon the islands was the property of one of the natives.

The same went for the supposed sexual permissiveness of the Tahitians. Polynesian sexual morality, far from permitting unbridled passion, was in truth much the same as that actually practised in England or France. True, Cook admitted, when his ship had first landed it was surrounded by loose women eager to sell their sexual favours. But would not a Tahitian find exactly the same practices if he paddled into Portsmouth or Chatham?

This vignette offers a fascinating window onto the play of Enlightenment values. The imaginative Parisian *philosophe*, Diderot, dragooned his Tahitians into service as a device for deriding the Catholic Church's morbid obsession with chastity and its killjoy attitudes towards sex. Captain Cook, the practical Englishman on the spot, would, by contrast, have no truck with such romances about 'noble savages'. Here, one might conclude, the *philosophes*, for all their fine talk about a 'science of man', seem caught red-handed spinning fantasies, whereas the no-nonsense English sailor was the true upholder of hard facts.

But the contrast just implied between Enlightenment ideology and sturdy empiricism is far too crude. For Cook's own perception of the Tahitians was itself profoundly shaped by theories and preconceptions of an enlightened nature. His 'defence' (as he saw it) of the islanders' sexual morality stemmed from the widely-held 'uniformitarian' and 'cosmopolitan' conviction – one endorsed by such *philosophes* as Voltaire and Hume – that human nature and behaviour were inevitably much the same the whole world over [139]. Moreover, Cook accepted the assumptions, spelt out most explicitly by the Scottish economists, that private property and social stratification were features intrinsic to any complex and flourishing society. Unlike the Australian aborigines, who sported hardly a stitch to hide their nakedness, the Tahitians formed a thriving people; therefore they must, thought Cook, have a system of social rank and private property [31; 116].

Possibly Diderot the armchair *philosophe* was more of a myth-maker than Cook the observer; but Cook himself also subscribed to fundamental Enlightenment values. Not for a moment would he interpret Polynesian sensuality as proof that those people were sunk in original sin. The mere fact that they possessed *different* customs and lifestyles from the Europeans did not automatically make them *inferior*, still less provide a justification for treating them unjustly, exploiting them, or selling them into slavery. For they were human, and Cook, along with all other *philosophes*, would undoubtedly have endorsed the dictum of Terence, the Roman playwright: '*homo sum, et nihil humanum alienum a me puto*' (I am a man, and I think nothing human alien to myself). Cook put into practice the Enlightenment maxim, best expressed by Montesquieu, that one should not sit in judgement upon the ways of other peoples, but rather seek to understand them in the context of their circumstances, and then use one's knowledge of them to improve understanding of oneself [18; 5].

This example has raised an important issue. How far should we see *all* the writings of the eighteenth century – *all* its works of art, science, and imagination – as expressions of the Enlightenment? Or would it be better to restrict that term to a much more self-contained body of polemics, addressed to the ends of criticism and reform? There is no single correct, definitive solution. We must explore how far Enlightenment attitudes permeated the entire culture of the times. The more we dilute such terms, however, the greater the danger of devaluing the currency, and ultimately producing meaninglessness [49].

Thus, it is important to note that Samuel Johnson's moral fable *Rasselas* (1759) shares many features with Voltaire's *Candide*, published in the same year. Through a picaresque narrative, both tell of the difficulties of finding true happiness in a world tormented by cruelty, violence, ambition, suffering and disappointment, one in which most supposedly 'happy' people – not least, princes and philosophers – are happy only because of their infinite talent for self-deception.

Voltaire's work is undoubtedly a key Enlightenment text, for it questions Divine benevolence, repudiates arid rationalism (so-called philosophical 'Optimism'), and urges practical activity [153]. It would hardly be accurate to see Johnson's parallel fiction as endorsing identical values. Johnson hated Voltaire, detested irreligion, and distrusted innovation. His own moral tale was a warning of the futility of over-inflated expectations of worldly bliss, not a handbook on how to be happy. *Rasselas* closed with a 'Conclusion in which Nothing is Concluded' [102].

In other words, it would be a mistake to label all intellectual developments, all innovations in literary forms, and all changes in aesthetic taste during the eighteenth century as expressions of a coherent Enlightenment philosophy. But it would be equally silly to deny that the notions of human nature and the ideals of the good life developed by the *philosophes* found wide expression in arts and letters, in print culture, and in practical life. In many respects the arts embodied the ideas of the Enlightenment.

Take the development of fiction. It cannot be said, with accuracy, that the Enlightenment gave birth to the novel. But many eighteenth-century novels explicitly addressed themselves to Enlightenment debates. Early in the century, Daniel Defoe's *Robinson Crusoe* classically fictionalized the dilemma of the man set in the state of nature – the shipwrecked mariner on the island – having to (re)invent civilization single-handedly (but for Man Friday) and create his own destiny. At the century's close, the pornographic novels of the Marquis de Sade addressed the problem of the rules of conduct (are there any at all?) in a post-Christian world in which good and evil, right and wrong, have collapsed into purely subjective questions of pleasure and pain [44; 136; 158].

Throughout the century, there was lively interaction between innovations in philosophy, morality and psychology on the one hand, and the exploration of character and motivation in fiction on the other; Diderot's dialogues, for instance *Rameau's Nephew*, offer good instances of this. Diderot used the dialogue form as a device for putting different points of view, for throwing discrete lights onto the same subject.

Similarly, the new respect paid to feeling and sentiment as the springs of true morality, by such thinkers as the 3rd Earl of Shaftesbury, Francis Hutcheson, and David Hume, rapidly found expression in novels such as Henry Mackenzie's *The Man of Feeling*, and in the works of the German writer Lessing [29]. Goethe's imaginative exploration, in his *Elective Affinities*, of the rival attractions of love, sex, and marital devotion explicitly drew upon the fashionable scientific theory of the differential attractiveness ('affinities') of distinct chemical elements. Laurence Sterne similarly told readers of his novel *Tristram Shandy* that they had to be familiar with Lockean psychology, for his *Essay Concerning Human Understanding* was the key to a proper grasp of the workings of the human mind [165].

Thus novels were often vehicles for exploring the implications of Enlightenment ideas. The same is true for many other genres of art and literature. Take opera. Not all developments within opera, of course, registered Enlightenment views or taste. Yet philosophical problems and persuasions certainly played their part. Mozart's *Don Giovanni* explores the tensions between libertine sexuality and the demands of humanity; his *The Marriage of Figaro* provocatively denounces the ancient 'feudal' *droit de seigneur*; *Il Seraglio* contrasts civilized Europe to exotic, yet barbarically despotic, Ottoman ways; while *The Magic Flute* presents the prospect of the spiritual improvement of human nature, realized through self-knowledge.

Not least, modern answers to the old questions: 'What is style?' and 'What is taste?' were developed by Enlightenment thinkers. The traditional conviction that the different genres of writing and art (epic, tragedy, mythological painting, etc.) and the canons of beauty governing them had been enunciated for all time by Aristotle and other Ancients was finally challenged and abandoned. Leading Enlightenment critics, from Shaftesbury to Diderot, Winckelmann to Lessing, instead sought to formulate a new philosophy of aesthetics. Their speculations demonstrated how taste was the product partly of cultural conditioning, and partly of psychological (and even physiological) responses to particular shapes, colours and sounds. The grandeur of an ocean or mountain, argued Edmund Burke, affected the senses, nerves and imagination in such a way as to create that feeling of awe experienced as the 'sublime' [76].

Beyond the arts, many other activities may likewise be seen to have been undergoing transformation thanks to a fruitful interplay between specific 'internal' pressures, and new views stimulated by Enlightenment values. Take medicine for instance. Peter Gay has rightly argued that one facet of the optimistic Enlightenment 'recovery of nerve' was that mankind was growing less fatalistic in the face of disease [61: vol. 2, ch. 1]. Fewer people resigned themselves to plagues and epidemics as being the will of God, just punishment for man's sins. Thus the physician and *philosophe* Erasmus Darwin deemed it absurd to suppose that the Devil caused disease, and grotesque to believe that God deliberately sent suffering – how could He be that cruel? [96].

When inoculation against smallpox became available in Britain early in the century, it was only certain fundamentalist Calvinist parishes in Scotland which resisted its introduction, on the grounds that affliction was predestined; enlightened England, with its more 'rational' belief that God helps those who help themselves, adopted that major medical advance more readily. Even bishops commended it, while the secretary of the Royal Society of London compiled statistics to show that the balance of probability was that inoculation saved lives. In a comparable way, Enlightenment doctors finally abandoned the traditional Biblical notion that mental illness was caused by diabolical possession. Instead they explained it either as a disease of the brain, or, drawing upon Locke's psychology, as mental delusion caused by the misassociation of ideas in the understanding. The 'old-fashioned' Methodist leader John Wesley, by contrast, clung to the belief that demons and witchcraft caused disease [123].

Indeed, the whole progression of life from the cradle to the grave underwent modification in the light of enlightened values. Take childbirth. Progressive doctors urged that 'ignorant' peasant midwives be abandoned in favour of expert male obstetricians for delivering babies. Once safely in this world, new-born infants, they argued, should no longer be subjected to the traditional custom of swaddling, but rather allowed to move 'naturally'. Young children should not be artificially 'coddled', but encouraged to exercise freely in the fresh air. That way, they would 'harden' and grow up strong [163]. If people attended to the health of their bodies (instead of wasting time on their 'souls'), there was no good reason, thought Erasmus Darwin and other like-minded physicians, why humans should not live well beyond the Biblical span of three-score years and ten.

But when death was finally approaching, Enlightenment thinkers believed it should be faced free of the traditional Christian horror of hellfire. When the Christian James Boswell, who was terrified of dying, visited the unbeliever David Hume, then dying of cancer, he was amazed, and angered, by his cheerful calm. Enlightenment thinkers held that dying was as natural as falling asleep [123; 95].

It would be facile, however, to give the impression that all the modifications the Enlightenment introduced into ways of living were unambiguous 'improvements'. The changes often involved mixed blessings. Take the treatment of criminals. The penalties for crime traditionally favoured by the *ancien régime* were corporal and capital punishment. In most Continental nations, though not

61

in Britain, judicial torture was still legal. Such Enlightenment spokesmen as Beccaria in Italy and Bentham in England condemned all such systems of punishment as inefficient as well as cruel. Bentham advocated lengthy jail sentences instead, during which the felon would work to repay society for his crimes. Reformers believed that, kept in his cell in solitary confinement, the criminal was bound to reflect upon his misdeeds, and would hence undergo psychological regeneration. Thus an enlightened penology would be less concerned with revenge and retribution than with deterrence and reform. Above all, judicial torture, which was always counterproductive, must be abolished. Such policies were widely implemented in Europe from the late eighteenth century onwards. Whether this new prison system has truly proved more humane, or even has been more successful in deterring crime, remains, however, open to question [56; 143].

Enlightenment reform did not always turn out as planned. Likewise, it was not always the men of Enlightenment who deserved the credit for true advances. Take slavery. As developed on colonial plantations, slavery was deplored by all the *philosophes*. It was a crime against humanity, it denied the brotherhood of man, and not least, claimed Adam Smith, it was actually more expensive than free wage-labour. Yet the campaigns in England which secured the abolition first of the slave trade and then of slavery itself in the British Empire were led by Evangelical Christians and Quakers. Thomas Jefferson, scion of Enlightenment, advocate of the rights of man, and third president of the United States, remained a slave-owner all his life [5].

Relations between principles and practice, attitudes and actions, are always complex. This chapter has not aimed to demonstrate that all changes in ways of living introduced during the eighteenth century followed from Enlightenment initiatives. Nor has it argued that those which did, were unalloyed improvements. What is beyond dispute, however, is that promoters of Enlightenment values believed that improvements to human life were possible and desirable. It was the duty of the present generation to make the world better for those to come. *Philosophes* contended that science, technology and industry should be harnessed to enable man to dominate nature. They insisted that the workings of society itself could be understood through economics, statistics, and what we now call sociology; and, once understood through these 'social sciences', social relations could be better organized and more rationally controlled. And above all, they believed that eyes should be trained not upon the past – the good old ways of the good old days – but upon the future. Progress was not inevitable; but at least it was within man's grasp [145].

8 Conclusion: Did the Enlightenment Matter?

Vociferous reactionary ideologues during the 1790s blamed all the evils of the French Revolution, as they saw them, upon the *philosophes*. For Burke, as for the Abbé Barruel, the 'illuminati' had been visionaries drunk upon reason; their speciously attractive, pseudo-humanitarian projects and facile rhetoric had charmed the impressionable, and fatally undermined the *status quo*. The antagonists of the Enlightenment could certainly point to erstwhile *philosophes* who were deeply caught up in French Revolutionary politics. When Condorcet died in the Terror, and the democratic author of *The Age of Reason*, Tom Paine, narrowly escaped with his life, it was easy to imply that radical chickens were finally coming home to roost.

It is by-and-large an idle business to blame or praise the *philosophes* for what happened in 1789 and beyond. In any case, almost all its leaders were by then dead, so we cannot divine their reactions. Erasmus Darwin and other liberal reformers hailed the dawn of the Revolution, but had lost sympathy with it by the time of the execution of Louis XVI and the excesses of the Terror [20; 93].

It is pertinent, however, to ask what the *philosophes* achieved in their own times. Noble measures of reform, stimulated by Enlightenment principles and sometimes guided by enlightened ministers, were passed in many nations: Joseph II's abolition of serfdom in the Austrian Empire is a signal instance. Turgot, a *philosophe*, was appointed by Louis XVI to resolve the crisis of the French finances. He failed; but so would anybody. In England, the new 'free trade' political economy of the Scottish school was applauded, and steadily introduced, by Pitt the Younger and his

64

followers. The utilitarian philosophical radicalism systematized by Jeremy Bentham left its undeniable imprint upon nineteenth-century administrative reform, above all, the radical overhaul of the Poor Law.

It may be hard to find measures advocated by Montesquieu, Voltaire, Diderot, d'Alembert, Mably or Morelly, which actually came to pass [93]. Yet that may say less about the irrelevance of the *philosophes*, than about the catastrophic failure of the French monarchy to put its own house in order. In any case, the leaders of the 'High Enlightenment' were not primarily trading in mundane political nostrums; they were more concerned with making palpable hits with their criticism, and with a far more sweeping and imaginative attempt to create a new, more humane, more scientific understanding of man as a social and natural being. They were concerned less with blueprints than with analysis, less with conclusions than with questions. What is the nature of man? What is the basis of morality? Is man a social being or not? Or, as Diderot's final play put the question to man, *Est-il Bon? Est-il Méchant?* [Is he good? Is he wicked?] Diderot's lifework was a whirlwind of questions, doubts and ambiguities which cry out to be labelled quintessentially 'modern'.

How do we know? What is right and wrong? Are we just machines, programmed by inheritance, anatomy or chemistry, or conditioned by the environment? Or, on the other hand, do we have free will? Or, perhaps, do we merely *think* we have free will? Where have we come from? Where are we going? All these questions were asked over and over again, sometimes playfully, sometimes philosophically. What is beyond doubt is that this programme of urgent and ceaseless inquiry into the nature of man and the springs of human action inaugurated by the Enlightenment amounts to a radical rejection of, or at least a distancing from, the standard teachings about man, his duties, and his destiny, which all the Christian churches had been imparting authoritatively, through their creeds and catechisms, down the centuries.

Historians differ as to how radical, how applicable, the explicit political programmes of the Enlightenment finally were. What seems clear is that its true radicalism lies in making a break with the Biblical, other-worldly framework for understanding man, society and nature, as revealed in the Scriptures, endorsed by the churches, rationalized in theology, and preached from the pulpit.

65

As late as the close of the seventeenth century, Bishop Bossuet, Europe's most eminent historian, could write what he called a *Universal History*, in which he saw the history of mankind opening less than six thousand years earlier, subordinated all human affairs to the Divine Will, and in the process, as Voltaire impishly noted, omitted the Chinese entirely [107]. For Christian history, the proper study for mankind was Providence. Philosophical history, as pioneered by Voltaire, took as its subject by contrast the actions of man in nature and society. Gibbon's *Decline and Fall of the Roman Empire* (1776–88) even offered a *natural* history of the Christian religion, interpreted as having progressed in the world from purely natural, or secondary, causes. *Philosophe* history – indeed, its perspectives upon man in general – replaced the divine frame of reference with the human.

The Enlightenment thus decisively launched the secularization of European thought. To say this, is not to claim that the *philosophes* were all atheists or that people thereafter ceased to be religious. Both are manifestly untrue. After all, the reaction against the French Revolution produced powerful evangelical and ecclesiastical revivals all over Europe. But, after the Enlightenment, the Christian religion ceased, once and for all, to preoccupy public culture. The Enlightenment is what sets Dante and Erasmus, Bernini, Pascal, Racine and Milton – all great Christian writers and artists – on one side of the great cultural divide, and Delacroix, Schopenhauer, George Eliot and Darwin on the other. Romanticism, one might suggest, is what is left of the soul when the religion has been drained out of it.

As the Enlightenment gained ground, it spelt the end of public wars of faith, put a stop to witch-persecutions and heretic-burnings, and signalled the demise of magic and astrology, the erosion of the occult, the waning of belief in the literal, physical existence of Heaven and Hell, in the Devil and all his disciples. The supernatural disappeared from public life. To fill the gap, nineteenth-century sentimentality had to endow Nature with its own holiness and invent new traditions, above all, a public show of patriotism. Religion remained, of course, but it gradually lost its props in learning, science, and in the well-stocked imagination. The Enlightenment sapped their credibility.

All such massive changes did not occur overnight. But they did happen. Why? Certain general forces were evidently at work,

66

for example the success of scientific inquiry. It would be grossly misleading to imply that the new science was involved in a pitched battle against religion. Far from it. Most eighteenth-century scientists were men of piety. But through the eighteenth century the discoveries of science, alongside other forms of investigation, were constantly undermining that unique sense of limited time and particular space which the Biblical story, with its Garden of Eden, and events in Bethlehem, and St Peter's successor installed in Rome, needed for its plausibility. Once intellectuals squarely faced the problems to belief posed by billions of stars occupying infinite space, and millions of years, and countless fossils of extinct creatures, and (no less) the history of man's linguistic, cultural and racial diversity throughout the five continents, Christianity was ever after with its back against the wall, forever trying to accommodate itself to new knowledge. The Enlightenment by contrast eagerly seized upon the excitement of the infinite.

It must above all be emphasized, however, as suggested above in Chapter 4, that the Enlightenment was the era which saw the emergence of a secular intelligentsia large and powerful enough for the first time to challenge the clergy. For centuries the priesthood had commanded the best broadcasting media (churches, pulpits), had monopolized posts in the leading educational establishments (schools, universities, seminaries), and had enjoyed legal privileges over the distribution of information.

This changed. It was during the eighteenth century that substantial bodies of *literati* outside the churches began to make a living out of knowledge and writing. Some earned a crust from Grub Street journalism; a few got very rich on the proceeds of their pens. Voltaire said that in his youth, society had been dominated by the well-born; later it had been taken over by men of letters. Such propagandists exploited such new channels of communication as newspapers and magazines (the *Spectator* might be called a kind of daily secular sermon) and utilized the opportunities offered by public opinion in what Habermas has dubbed the 'public sphere' [69].

They appealed to an ever-broadening reading public, eager for new forms of writing, such as essays and fiction and biography. In turn, their impact was reinforced by such secular institutions as the reading clubs, academies, and literary and scientific societies mentioned above. The First Estate, the 'Lords

Spiritual', was thus challenged by a new body, the 'Fourth Estate' (roughly, the press), in a struggle to win the ear of the 'Second Estate' (the traditional political classes) and the emergent 'Third Estate' (the Commons).

It is not easy to make a balanced assessment of the significance of this shift produced by the Enlightenment through the emergence of a powerful body of *literati*, in other words, the intelligentsia (or what Coleridge neatly called the 'clerisy'). In the early modern centuries when knowledge and opinion were still largely expounded by clergymen who owed formal loyalty to higher authorities, the production of ideas followed orthodox and predictable patterns, as was perhaps appropriate for a relatively stable traditional society. The new intelligentsia by contrast had loyalties which were infinitely more varied. Sometimes they wrote for patrons, or for paymasters. But often they wrote to please themselves, or with a broad sense of communicating to a general paying 'public' out there. And, as writers freed themselves from the fetters which had constrained the clergy, the world of letters became deeply diversified. The multiplication of mouthpieces, each clamouring for attention, matched and enhanced the growing diversity of articulate society at large, as literacy rates improved and more people read pamphlets and newspapers [161].

We might suggest this meant greater independence, for writers and readers alike. Or, in other words, the legacy of the Enlightenment was thus the emancipation of the European mind from the blinkers of dogma. If so, the ultimate impact of the Enlightenment would best be characterized as radical. Yet this is too simple, and doubts arise. Ideas never run far ahead of society. And so much of the daring, innovative thought of the eighteenth century was quickly recycled to become the stock props of the established order of the nineteenth.

The brave new Enlightenment sciences of man – analysing social dynamics, population growth, and wealth creation – became the positivistic 'dismal sciences' which were soon to provide perfect ideological fodder for governments eager to explain why capitalist relations were immutable and ineluctable, why poverty was the fault of the poor. The challenging psychology of Condillac and Helvétius, which stated that human beings were pregnant with possibilities, was readily co-opted to ensure obedience and discipline amongst children at school and adults in the workplace.

What had once been the exciting vision of 'man the machine' (free of original sin) became the nightmare reality of factory life in the machine age – or, later, behaviourist conditioning [65].

The Enlightenment helped to free man from his past. In so doing, it failed to prevent the construction of future captivities. We are still trying to solve the problems of the modern, urban industrial society to which the Enlightenment was midwife. And in our attempts to do so, we largely draw upon the techniques of social analysis, the humanistic values, and the scientific expertise which the *philosophes* generated. We remain today the Enlightenment's children.

Reading Suggestions

The following bibliography lists the tiniest fraction only of scholarship on the Enlightenment. My principle of selection has been to concentrate heavily upon works most likely to be readily available and profitable to college students, undergraduates and sixth formers. I have included reliable standard accounts, some of them now old but at least widely available in libraries, as well as new interpretations. For practical reasons, where topics are adequately covered by books in the English language, I have cited these in preference to works in French, German, Italian, etc. For similar reasons of accessibility, I have also concentrated upon books rather than articles.

For those with good linguistic skills, and/or access to research libraries, there are excellent bibliographical guides which will provide entrée into more specialized literature than I have opted to list here. *The Eighteenth Century: A Current Bibliography* is published every year by the American Society for Eighteenth Century Studies. It contains a magisterial listing, and in some cases critical evaluation, of almost everything falling within the field of Enlightenment studies (annual volumes of this publication currently run to over 500 pages!). Before 1975, this bibliography was produced as an annual *Supplement* to the journal *Philological Quarterly*. Since the mid-1950s, the Voltaire Foundation has produced many volumes a year in a series entitled *Studies on Voltaire and the Eighteenth Century*. A complete listing, volume by volume, is available from the Foundation at the Taylor Institute, Oxford. Amongst these volumes are the full published proceedings of numerous International Conferences on the History of the Enlightenment; these present a fascinating profile of the changing face of research and interpretation.

Additionally, numerous journals publish extensively in Enlightenment studies, and review books in the field. Notable amongst these are *Eighteenth Century Studies*, *Eighteenth Century Life*, *The Eighteenth Century: Theory and Interpretation* (formerly known as *Studies in Burke and his Times*), the *Journal of the History of Ideas*, *Enlightenment and Dissent*, *Studies in Eighteenth Century Culture*, and the *The British Journal for Eighteenth Century Studies*. Specialist publications exist for many leading *philosophes*, e.g. *Diderot Studies*.

Furthermore, Peter Gay's *The Enlightenment: An Interpretation* [61] contains a superbly detailed critical bibliography of works which appeared up to the mid- to late 1960s. For more modern bibliographical guidance, the *Blackwell Companion to the Enlightenment* is also useful. I have chosen not to detail below modern scholarly editions of the works or letters of the leading *philosophes*. The first part of this bibliography, however, lists anthologies of extracts, in English translation where appropriate, from the writings of the leading lights of the movement.

Section A   Reference and Anthologies

[1] Berlin, Isaiah (ed.), *The Age of Enlightenment* (New York: Mentor, 1956). Berlin's selection of texts focuses primarily upon philosophical issues and authors.

[2] Black, Jeremy, and Roy Porter (eds), *The Basil Blackwell Dictionary of World Eighteenth-Century History* (Oxford: Blackwell, 1994).

[3] Brinton, Crane (ed.), *The Portable Age of Reason Reader* (New York: Viking, 1956).

[4] Alexander Broadie (ed.), *The Scottish Enlightenment: An Anthology* (Edinburgh: Canongate, 1997). A generous selection.

[5] Carretta, Vincent (ed.), *Unchained Voices: An Anthology of Black Authors in the English-Speaking World of the Eighteenth Century* (Lexington, KY: University Press of Kentucky, 1996). The range of Black authors in the Enlightenment period has only recently come to light.

[6] Crocker, Lester G. (ed.), *The Age of Enlightenment* (New York: Harper, 1969).

[7] Eliot, Simon, and Beverley Stern (eds), *The Age of Enlightenment*, 2 vols (New York: Barnes and Noble, 1979). One of the most comprehensive anthologies, with extensive space given to science and the arts, though with some eccentric omissions, e.g. Rousseau.

[8] Gay, Peter, *The Enlightenment* (New York: Simon and Schuster, 1973). A very generous and balanced collection.

[9] ——, *Deism: An Anthology* (Princeton, NJ: Princeton University Press, 1968). An eye-opening selection of mainly early Enlightenment religious writings.

[10] Kramnick, Isaac (ed.), *The Portable Enlightenment Reader* (Harmondsworth: Penguin, 1995).

[11] Marsak, L. (ed.), *The Enlightenment* (New York: Wiley, 1972).

[12] Rendall, Jane (ed.), *The Origins of the Scottish Enlightenment, 1707–76* (London: 1978). A useful collection of documents on the Scottish Enlightenment.

[13] Schmidt, James (ed.), *What is Enlightenment? Eighteenth-Century Answers and Twentieth-Century Questions* (Berkeley, CA: University of California Press, 1996). Reproduces eighteenth-century contributions to the 'What is Enlightenment?' debate.

[14] Williams, David (ed.), *The Enlightenment* (Cambridge: Cambridge University Press, 1999). Mainly political theory.

[15] Yolton, John W. (ed.), *The Blackwell Companion to the Enlightenment* (Oxford: Blackwell, 1991). The best work of reference.

Section B   Interpretative Works

[16] Anchor, Robert, *The Enlightenment Tradition* (Berkeley, CA: University of California Press, 1967). An enthusiastic, if flawed, attempt to explore the Enlightenment as an expression of bourgeois values.

[17] Anderson, Douglas, *The Radical Enlightenment of Benjamin Franklin* (Baltimore, MD: Johns Hopkins University Press, 1997). Shows how Franklin's early reading made him a man of the Enlightenment.

[18] Aron, Raymond, *Main Currents in Sociological Thought* (London: Penguin, 1968). Contains an important discussion of Montesquieu as one of the founding fathers of the social sciences.

[19] Baker, K. M., *Condorcet: From Natural Philosophy to Social Mathematics* (Chicago: Chicago University Press, 1975). A fine intellectual biography of the late Enlightenment's most important social scientist.

[20] ——, 'Enlightenment and Revolution in France: Old Problems and Renewed Approaches', *Journal of Modern History*, 53 (1981): 281–303. Argues that we will not be able properly to assess the impact of the Enlightenment upon the French Revolution until we know the history of the late *ancien régime* much better.

[21] Becker, Carl, *The Heavenly City of the Eighteenth-Century Philosophers* (New Haven: Yale University Press, 1932). Becker provocatively claims that the *philosophes'* demolition of Christianity involved an equal investment of faith in the religion of reason. The *philosophes* created as many myths as they destroyed.

[22] Behrens, C. B. A., *Society, Government and the Enlightenment: The Experiences of Eighteenth-Century France and Prussia* (London: Thames and Hudson, 1985). Important comparative survey of the ideals and effectiveness of 'enlightened absolutism'.

[23] Bene, E., and I. Kovacs (eds), *Les Lumières en Hongrie, en Europe Centrale et en Europe Orientale* (Budapest, 1975). Essays on the

impact of the Enlightenment on the feudal regimes of central and eastern Europe.

[24] Berry, Christopher J., *The Idea of Luxury: A Conceptual and Historical Investigation* (Cambridge: Cambridge University Press, 1994). The status of luxury (vice or virtue?) was hotly contested throughout the Enlightenment.

[25] Berry, Christopher J., *Social Theory of the Scottish Enlightenment* (Edinburgh: Edinburgh University Press, 1997). More up-to-date interpretations, if also more jargon-laden, than in Bryson [31].

[26] Blanning, T. C. W., *Reform and Revolution in Mainz, 1743–1803* (Cambridge: Cambridge University Press, 1974). Shows the importance of Enlightenment ideas in promoting bureaucratic reform from within the German principalities.

[27] ——, *The French Revolution: Aristocrats versus Bourgeois?* (London: Macmillan, 1987). Airs, and scotches, the notion that the French Revolution was a bourgeois revolution.

[28] Brown, S. C. (ed.), *Philosophers of the Enlightenment* (Brighton: Harvester Press, 1979). Eleven lucid essays by distinguished authorities examining the philosophies of the main *philosophes* from Locke to Kant.

[29] Bruford, W. H., *Germany in the Eighteenth Century: The Social Background of the Liberal Revival* (Cambridge: Cambridge University Press, 1952). Strong on the development of the cultural ideals associated with the German Enlightenment.

[30] Brumfitt, J. H., *Voltaire, Historian* (Oxford: Oxford University Press, 1958). Rightly stresses Voltaire's significance in the development of philosophical attitudes towards the past and in pioneering the cultivation of social history.

[31] Bryson, Gladys, *Man and Society: The Scottish Inquiry of the Eighteenth Century* (Princeton, NJ: Princeton University Press, 1945). Though dated, still a fine general survey of Scottish social thought in the Enlightenment, rightly emphasizing the importance of 'conjectural history'.

[32] Burckhardt, Jakob, *The Civilization of the Renaissance in Italy* (Oxford: Phaidon, 1981). The classic statement of the case for the 'discovery' of 'man' in the fifteenth-century Italian Renaissance.

[33] Burke, Peter, *The Renaissance Sense of the Past* (New York: St Martin's Press, 1970). Examines the growing sense of history (above all, the discovery of 'anachronism') in early modern Europe.

[34] Byrne, James, *Glory, Jest and Riddle: Religious Thought in the Enlightenment* (London: SCM Press, 1996). A brief, lively Europe-wide survey.

[35] Cassirer, Ernst, *The Philosophy of the Enlightenment* (Boston, MA: Beacon, 1964). One of the first modern works to take Enlightenment philosophy seriously, Cassirer's is still the leading account of the metaphysics underlying eighteenth-century thought.

[36] Champion, Justin A. I., *The Pillars of Priestcraft Shaken: The Church of England and its Enemies, 1660–1730* (Cambridge: Cambridge University Press, 1992). The best account of the English Deists.

[37]   Chartier, Roger, *Cultural History: Between Practices and Representations*
       (Ithaca, NY: Cornell University Press, 1988). Chartier explores the
       complex texture and meanings of 'culture' for the eighteenth
       century. Some of his essays address the problem of how far
       Enlightenment ideas percolated down into the reading matter of
       the peasantry.
[38]   Chitnis, Anand, *The Scottish Enlightenment: A Social History* (London:
       Croom Helm, 1976). Examines the relations between socio-
       economic change and Scottish Enlightenment social theory.
[39]   Claeys, Gregory, *Thomas Paine: Social and Political Thought* (Winches-
       ter, MA: Unwin Hyman, 1989). Strong on historical context.
[40]   Clark, William, Jan Golinski and Simon Schaffer (eds), *The Sciences
       in Enlightened Europe* (Chicago: University of Chicago Press, 1999).
       Contains original and interpretative essays. Use in conjunction
       with [121].
[41]   Cohen, Estelle, ' "What the Women at all Times Would Laugh At":
       Redefining Equality and Difference, circa 1660–1760', *Osiris*, XII
       (1997): 121–42.
[42]   Cranston, Maurice, *Jean-Jacques: The Early Life and Work of Jean-
       Jacques Rousseau* (London: Allen Lane, 1983). An up-to-date and
       reliable biography.
[43]   ——, *Philosophers and Pamphleteers: Political Theorists of the Enlighten-
       ment* (Oxford: Oxford University Press, 1986). Cranston covers the
       main French theorists from Montesquieu to Condorcet, showing
       the emergence of more 'populist' ideas in the generation of the
       French Revolution.
[44]   Crocker, Lester G., *An Age of Crisis: Man and World in Eighteenth
       Century France* (Baltimore, MD: Johns Hopkins University Press,
       1959).
[45]   ——, *Nature and Culture: Ethical Thought in the French Enlightenment*
       (Baltimore, MD: Johns Hopkins University Press, 1963).
          In both these works, Crocker highlights the dilemmas produced
       by Enlightenment naturalism, subjectivism and relativism, and its
       critique of the traditional Christian basis of moral and social
       values.
[46]   Darnton, Robert, *The Business of Enlightenment: A Publishing
       History of the Encyclopédie, 1775–1800* (Cambridge, MA: Harvard
       University Press, 1979). A major account of the production
       and distribution of the *Encyclopédie*, emphasizing the role of the
       book-trade.
[47]   ——, *The Literary Underground of the Old Regime* (Cambridge, MA:
       Harvard University Press, 1982). Gathers together Darnton's
       important essays upon the popularization of Enlightenment ideas
       through the press, booksellers etc., and emphasizes the distinc-
       tions between 'high' and 'low' culture.
[48]   ——, *The Forbidden Best-Sellers of Pre-Revolutionary France* (London:
       HarperCollins, 1996). A major discussion of French erotica as
       embodiments of enlightened views.

[49]   ——, 'George Washington's False Teeth', *The New York Review of Books*, 27 March 1997. Darnton takes issue with the postmodernist attack on the Enlightenment.

[50]   Davis, Natalie Zemon, and Arlette Farge (eds), *A History of Women in the West*, vol. iii: *Renaissance and Enlightenment Paradoxes* (Cambridge, MA: Harvard University Press, 1993).

[51]   Dinwiddy, John, *Bentham* (Oxford: Oxford University Press, 1989). A clear survey of Bentham as both thinker and activist.

[52]   Doyle, W., *The Ancien Régime* (London: Macmillan, 1986). The best brief modern reassessment of the socio-political order criticized by the *philosophes*.

[53]   Dülmen, Richard van, *The Society of the Enlightenment: The Rise of the Middle Class and Enlightenment Culture in Germany*, trans. Anthony Williams (Cambridge: Polity Press, 1992). Emphasizes the role of reading and debating societies as vehicles of enlightened thinking.

[54]   Eisenstein, Elizabeth, 'On Revolution and the Printed Word', in Roy Porter and Mikuláš Teich (eds), *Revolution in History* (Cambridge: Cambridge University Press, 1986), pp. 186–205. Argues for the prime importance of the press and the book-trade for the spread of radical ideas in the Enlightenment, and discusses their significance for the Revolution.

[55]   Foucault, Michel, *Madness and Civilization: A History of Insanity in the Age of Reason* (New York: Pantheon, 1965). Foucault argues that the 'Age of Reason' showed no tolerance towards 'unreason', or insanity. The Enlightenment resulted in the insane being confined rather than freed.

[56]   ——, *Discipline and Punish* (London: Allen Lane, 1977). A powerful account of the darker implications of supposedly enlightened attitudes towards punishment.

[57]   Fox, Christopher, Roy Porter and Robert Wokler (eds), *Inventing Human Science: Eighteenth Century Domains* (Berkeley, CA: University of California Press, 1995). Essays on the early history of the various social sciences.

[58]   Gascoigne, John, *Joseph Banks and the English Enlightenment: Useful Knowledge and Polite Culture* (Cambridge and New York: Cambridge University Press, 1994). Brings out the various meanings of science for a man of the Enlightenment.

[59]   Gay, Peter, *The Party of Humanity: Essays in the French Enlightenment* (New York: Norton, 1971).

[60]   ——, *Voltaire's Politics: The Poet as Realist* (New York: Vintage, 1956).
       In these two works, Gay above all convincingly rescues Voltaire's reputation as a skilled and principled political activist and propagandist.

[61]   ——, *The Enlightenment: An Interpretation*, 2 vols (New York: Vintage, 1966–9). Gay offers the best, modern sympathetic account of the Enlightenment as the source of the modern liberal humanism which is the basis of our own values.

[62] Gijswijt-Hofstra, Marijke, Brian P. Levack and Roy Porter, *Witch-craft and Magic in Europe*, vol. 5: *The Eighteenth and Nineteenth Centuries* (London: Athlone, 1999). Gives equal weight to the legal, social and intellectual dimensions.

[63] Goldmann, Lucien, *The Philosophy of the Enlightenment: The Christian Burgess and the Enlightenment* (Cambridge, MA: Harvard University Press, 1973). Challenging Marxist account which emphasizes the ambiguities and paradoxes of the Enlightenment for the bourgeoisie.

[64] Goodman, Dena, *The Republic of Letters: A Cultural History of the French Enlightenment* (Ithaca, NY, and London: Cornell University Press, 1994). Goodman particularly emphasizes the role of the salon.

[65] Gray, John, *Enlightenment's Wake: Politics and Culture at the Close of the Modern Age* (London: Routledge, 1995).

[66] Grell, Peter, and Roy Porter (eds), *Toleration in the Enlightenment* (Cambridge: Cambridge University Press, 2000).

[67] Grimsley, Ronald, *Jean d'Alembert, 1717–83* (Oxford: Clarendon Press, 1963). A lucid life of Diderot's partner on the *Encyclopédie*, a mathematician of some importance in the history of science.

[68] ——, *The Philosophy of Rousseau* (Oxford: Oxford University Press, 1973). Grimsley emphasizes the religious and pre-romantic underpinnings of Rousseau's thought.

[69] Habermas, Jürgen, *The Structural Transformation of the Public Sphere: An Inquiry into a Category of Bourgeois Society*, trans. Thomas Burger (Cambridge: Polity, 1989); originally published as *Strukturwandel der Öffentlicheit* (Berlin: Luchterhand, 1962).

[70] Hacking, Ian, *The Taming of Chance* (Cambridge: Cambridge University Press, 1990).

[71] Halévy, Elie, *The Growth of Philosophic Radicalism* (London: Faber and Faber, 1928). Still the best history of utilitarian thinking. Halévy concentrates upon Britain but pays some attention to French thinkers.

[72] Hampson, Norman, *The Enlightenment* (Harmondsworth: Penguin Books, 1968). Probably the liveliest modern single-volume survey.

[73] Hankins, Thomas L., *Science and the Enlightenment* (Cambridge and New York: Cambridge University Press, 1985).

[74] Hazard, Paul, *The European Mind, 1680–1715* (Cleveland, OH: Meridian, 1963).

[75] ——, *European Thought in the Eighteenth Century: From Montesquieu to Lessing* (Cleveland, OH: Meridian, 1963).
In these two books, Hazard captures, in vivid language, the effervescence of intellectual transformation produced throughout the eighteenth century by the new science, scholarship and geographical discoveries of the seventeenth.

[76] Hipple, Walter John, *The Beautiful, the Sublime and the Picturesque in Eighteenth Century British Aesthetic Theory* (Carbondale, IL: Southern Illinois University Press, 1957). A major interpretation of the new

aesthetics of the eighteenth century and its philosophical foundations.

[77]   Herr, Richard, *The Eighteenth Century Revolution in Spain* (Princeton, NJ: Princeton University Press, 1958). Far the best account of the (rather limited) impact of the Enlightenment in Spain.

[78]   Hoffmann, Paul, *La Femme dans la pensée des lumières* (Paris: Ophrys, 1977). The fullest survey of the (often rather contradictory) attitudes of *philosophes* to the nature of woman and her place in society.

[79]   Hont, I., and M. Ignatieff (eds), *Wealth and Virtue: The Shaping of Political Economy in the Scottish Enlightenment* (Cambridge: Cambridge University Press, 1983). A collection of essays which situate the rise of modern Smithian political economy in the traditions of morality and natural law prevalent in Scotland.

[80]   Horkheimer, Max, and Theodor W. Adorno, *Dialectic of Enlightenment* (New York: Herder and Herder, 1972). A critique of the Enlightenment from the viewpoint of the critical, Frankfurt school of philosophy, arguing that its much-vaunted scientific 'reason' proved not liberating but authoritarian.

[81]   Hundert, E. G., *The Enlightenment's Fable: Bernard Mandeville and the Discovery of Society* (Cambridge: Cambridge University Press, 1994).

[82]   Hunt, M., et al. (eds), *Women and the Enlightenment* (New York: Haworth Press, 1984). Pioneering studies. Use in conjunction with Hoffmann [78].

[83]   Im Hof, Ulrich, *The Enlightenment*, trans. William E. Yuill (Oxford: Blackwell, 1994).

[84]   Jacob, Margaret C., *The Radical Enlightenment: Pantheists, Freemasons and Republicans* (London: George Allen and Unwin, 1981). A challenging work which argues for the existence of a radical Enlightenment before the 'classic' Enlightenment associated with Montesquieu and Voltaire. Some of Jacob's conclusions have been hotly disputed.

[85]   ——, *Living the Enlightenment: Freemasonry and Politics in 18th Century Europe* (New York: Oxford University Press, 1992).

[86]   ——, *Scientific Culture and the Making of the Industrial West* (Oxford: Oxford University Press, 1997).

[87]   Kors, A. C., *D'Holbach's Circle: An Enlightenment in Paris* (Princeton, NJ: Princeton University Press, 1977). Kors lays bare the apparent paradox of the most radical atheistic opinions being entertained by the most elite of Paris circles.

[88]   ——, and Paul J. Korshin (eds), *Anticipations of the Enlightenment in England, France and Germany* (Philadelphia: University of Pennsylvania Press, 1987). Eleven up-to-date essays addressing aspects of the problem of the origins of Enlightenment, literary, philosophical and cultural.

[89]   Krieger, Leonard, *Kings and Philosophers, 1689–1789* (New York: Norton, 1970). The best study of the often fraught

interaction between Enlightenment thinkers and 'enlightened absolutism'.

[90] Levine, Joseph M., *The Battle of the Books: History and Literature in the Augustan Age* (Ithaca, NY: Cornell University Press, 1992). Traces the 'Ancients versus Moderns' struggle.

[91] Lindemann, Mary, *Health and Healing in Seventeenth- and Eighteenth-Century Germany* (Baltimore, MD: Johns Hopkins University Press, 1996). An account of the organization of medicine in German-speaking Europe.

[92] Lough, John, *Essays on the Encyclopédie of Diderot and D'Alembert* (London: Oxford University Press, 1968). Detailed and highly informative on both authorship and contents.

[93] ——, *The Philosophes and Post-Revolutionary France* (Oxford: Clarendon Press, 1982). Examines the reform programmes of the *philosophes* and assesses how far the French Revolution realized them.

[94] Mackie, Erin, *Market à la Mode: Fashion, Commodity, and Gender in The Tatler and The Spectator* (Baltimore, MD: John Hopkins University Press, 1997).

[95] McManners, J., *Death and the Enlightenment: Changing Attitudes to Death among Christians and Unbelievers in Eighteenth Century France* (Oxford: Clarendon Press, 1981). A fundamentally important and often moving account of the relationship between old Christian and newer Enlightenment attitudes, beliefs and practice.

[96] MacNeil, Maureen, *Under the Banner of Science: Erasmus Darwin and his Age* (Manchester: Manchester University Press, 1987). Examines Darwin not merely as a scientist (one of the earliest advocates of biological evolutionism) but as a spokesman for the emergent industrial bourgeoisie.

[97] Macpherson, C. B., *The Political Theory of Possessive Individualism* (Oxford: Oxford University Press, 1983). An analysis of the rise of political liberalism from a Marxist angle.

[98] Manuel, Frank E., *The Eighteenth Century Confronts the Gods* (New York: Atheneum, 1967). A stimulating survey of Enlightenment attempts to demystify religion.

[99] Marshall, John, *John Locke: Resistance, Religion and Responsibility* (Cambridge: Cambridge University Press, 1994).

[100] Mason, Haydn, *Voltaire: A Biography* (Baltimore, MD: Johns Hopkins University Press, 1981). The best modern life.

[101] —— (ed.), *The Darnton Debate: Books and Revolution in the Eighteenth Century* (Oxford: Voltaire Foundation, 1998). Essays by various scholars assessing Darnton's account of the political subversiveness of 'Grub Street' culture in the lead-up to the French Revolution.

[102] Mauzi, Robert, *L'Idée du bonheur dans la littérature et la pensée française au XVIII siècle* (Paris: Colin, 1960). Mauzi explores the implications for psychology and morality of the new Enlightenment

commitment to the pursuit of happiness as the goal of human life.

[103] May, Henry F., *The Enlightenment in America* (New York: Oxford University Press, 1976). The best survey, stressing the diversity of Enlightenment currents in the Colonies, and correctly emphasizing the role of liberal Protestantism in Enlightenment.

[104] Meek, Ronald (ed.), *The Economics of Physiocracy* (Cambridge, MA: Harvard University Press, 1962). Essays on the intellectual foundations of the movement in French economic thinking which saw the source of all value lying in the land and agriculture.

[105] Mornet, D., *Les origines intellectuelles de la révolution français 1715–1787* (Paris: Colin, 1932). Pathbreaking attempt to assess the impact of the Enlightenment upon pre-revolutionary French culture. Mornet cast doubt upon the common assumption that Voltaire, Rousseau etc., were amongst the most widely-read eighteenth-century authors.

[106] Myers, Sylvia Harcstark, *The Bluestocking Circle: Women, Friendship, and the Life of the Mind in Eighteenth-Century England* (Oxford: Clarendon Press, 1990). Surveys those British women of letters who hoped to raise their own status through the life of the mind.

[107] O'Brien, Karen, *Narratives of Enlightenment: Cosmopolitan History from Voltaire to Gibbon* (Cambridge: Cambridge University Press, 1997). Brings out the break with traditional Christian, Biblical universal history.

[108] Olson, Richard, *The Emergence of the Social Sciences, 1642–1792* (New York: Twayne, 1993). Comprehensive and clear.

[109] Outram, Dorinda, *The Enlightenment* (Cambridge: Cambridge University Press, 1995). By far the best modern discussion of recent historiographic trends. A valuable corrective to Gay [61].

[110] Oz-Salzberger, Fania, *Translating the Enlightenment: Scottish Civic Discourse in Eighteenth Century Germany* (Oxford: Clarendon Press, 1995). A helpful study in the cross-cultural transmission of ideas.

[111] Palmer, Robert R., *Catholics and Unbelievers in Eighteenth Century France* (New York: Cooper Square, 1961). Shows that the *philosophes'* picture of Catholic, and in particular Jesuit, intolerance and bigotry was largely a caricature.

[112] ——, *The Age of the Democratic Revolution: A Political History of Europe and America, 1760–1800*, 2 vols (Princeton, NJ: Princeton University Press, 1959–64). Palmer argues for a cumulative political transformation in America and Europe, based upon new democratic thinking and the importance of the example of the American War of Independence.

[113] Payne, H. C., *The Philosophes and the People* (New Haven, CT: Yale University Press, 1976). A perceptive account of the limits of the sympathies towards the people expressed by the French *lumières*. Good on the role of education.

[114] Phillipson, Nicholas, 'Adam Smith as Civic Moralist', in Istvan Hont and Michael Ignatieff (eds), *Wealth and Virtue: The Shaping of Political Economy in the Scottish Enlightenment* (Cambridge and New York: Cambridge University Press, 1983), pp. 179–202. Insightful account of Smith's theory of 'sympathy' as the model of a new social morality.

[115] ——, *Hume* (London: Weidenfeld and Nicolson, 1989). Brief but best.

[116] Pocock, J. G. A., *The Machiavellian Moment: Florentine Political Thought and the Atlantic Republican Tradition* (Princeton, NJ: Princeton University Press, 1975). An illuminating analysis of 'civic humanism' as a political tradition which stressed the virtues of small republics with high levels of citizen participation.

[117] ——, 'Post-Puritan England and the Problem of the Enlightenment', in P. Zagorin (ed.), *Culture and Politics from Puritanism to the Enlightenment* (Berkeley, CA: University of California Press, 1980), pp. 91–111. Develops Pococks ideas on the uniqueness of the Enlightenment in Britain.

[118] ——, *Barbarism and Religion*, 2 vols; vol. i: *The Enlightenments of Edward Gibbon, 1737–1764*; vol. ii: *Narratives of Civil Government* (Cambridge: Cambridge University Press, 1999). A rich account of the many contexts to Gibbon's *Decline and Fall*.

[119] Porter, Roy, *Edward Gibbon: Making History* (London: Weidenfeld, 1988). Sets Gibbon in the context of Enlightenment history-writing.

[120] ——, 'The Exotic as Erotic: Captain Cook at Tahiti', in G. S. Rousseau and Roy Porter (eds), *Exoticism in the Enlightenment* (Manchester: Manchester University Press, 1989), pp. 117–44.

[121] —— (ed.), *The Cambridge History of Science*, vol. 4: *The Eighteenth Century* (Cambridge: Cambridge University Press, 2000). A thorough survey of the entire range of enlightened sciences.

[122] ——, *Enlightenment: Britain and the Creation of the Modern World* (Harmondsworth: Penguin, 2000).

[123] ——, and Dorothy Porter, *In Sickness and In Health: The British Experience, 1650–1850* (London: Fourth Estate, 1988). Surveys changing attitudes and experiences about health and sickness, life and death.

[124] ——, and Mikuláš Teich (eds), *The Enlightenment in National Context* (Cambridge: Cambridge University Press, 1981). A series of essays examining the distinctive nature of the Enlightenment in different nations. Porter writes on England, Nicholas Phillipson on Scotland, Norman Hampson on France, Simon Schama on the Netherlands, Samuel S. B. Taylor on Switzerland, Owen Chadwick on Italy, Joachim Whaley on Protestant Germany, T. C. W. Blanning on Catholic Germany, Ernst Wangermann on Austria, Mikuláš Teich on Bohemia, Tore Frängsmyr on Sweden, Paul Dukes on Russia, and J. R. Pole on America.

[125] Proust, Jacques, *Diderot et l'Encyclopédie* (Paris: Colin, 1962). To be used in conjunction with Darnton [46].

[126] Racevskis, Karlis, *Postmodernism and the Search for Enlightenment* (Charlottesville: University Press of Virginia, 1993). An analysis of postmodernist critiques of Enlightenment claims to have privileged access to truth.

[127] Raeff, Michael, *The Well-Ordered Police State: Social and Institutional Change through Law in the Germanies and Russia, 600–1800* (New Haven, CT: Yale University Press, 1984). The soundest study of the philosophy of enlightened government in the absolutist states.

[128] Redwood, John, *Reason, Ridicule and Religion: The Age of Enlightenment in England* (London: Thames and Hudson, 1976; reprinted 1996). Though flawed and sometimes inaccurate, this remains the most up-to-date account of intellectual ferment in England.

[129] Rennie, Neil, *Far-fetched Facts: The Literature of Travel and the Idea of the South Seas* (Oxford: Clarendon Press, 1995). Explores the relations of fact and fantasy in enlightened travel writing.

[130] Roche, Daniel, *Les Républicains des lettres. Gens de culture et lumières au XVIII siècle* (Paris: Fayard, 1988). Roche uses statistical methods to examine the spread of Enlightenment ideas amongst the educated classes in the French provinces, and to document reading habits.

[131] ——, *France in the Enlightenment* (Cambridge, MA: Harvard University Press, 1998). The best survey of the socio-cultural history of eighteenth-century France.

[132] Roger, Jacques, *Les Sciences de la vie dans la pensée française au XVIII siècle* (Paris: Colin, 1963). By far the best introduction to the milieu of French scientific thought in the Enlightenment.

[133] Ross, Ian Simpson, *The Life of Adam Smith* (Oxford: Clarendon Press, 1995). Demonstrates how Smith was far more than an economist.

[134] Rossi, P. *The Dark Abyss of Time: The History of the Earth and the History of Nations from Hooke to Vico* (Chicago: University of Chicago Press, 1984). The best account of the Enlightenment discovery of the vast age of the universe and of the *philosophes*' attitudes towards Antiquity.

[135] Rousseau, G. S., and Roy Porter (eds), *The Ferment of Knowledge: Studies in the Historiography of Eighteenth Century Science* (Cambridge: Cambridge University Press, 1980). A survey of developments in science in the Enlightenment.

[136] ——, and —— (eds), *Sexual Underworlds of the Enlightenment* (Manchester: Manchester University Press, 1988). The essays explore the ambiguities of sexual liberty as proclaimed by the thinkers and literary propagandists of the Enlightenment.

[137] Schama, Simon, *The Embarrassment of Riches: An Interpretation of Dutch Culture in the Golden Age* (New York: Knopf, 1987). Schama

emphasizes the unique 'modernity' of the Dutch experience in the seventeenth century.

[138] Schiebinger, Londa, *The Mind Has No Sex? Women in the Origins of Modern Science* (Cambridge, MA: Harvard University Press, 1989).

[139] Schlereth, Thomas, *The Cosmopolitan Ideal in Enlightenment Thought* (Notre Dame, IN: University of Notre Dame Press, 1977). An able survey of Enlightenment 'universalism', the desire to transcend national boundaries and narrow parochialism.

[140] Schofield, R. E., *The Lunar Society of Birmingham* (Oxford: Oxford University Press, 1963). The best study of the interplay between an English coterie of *philosophes*, manufacturers and scientists.

[141] Schouls, Peter, *Reasoned Freedom: John Locke and Enlightenment* (Ithaca, NY: Cornell University Press, 1992). A clear and capable exposition of Locke's philosophy.

[142] Shackleton, Robert, *Montesquieu: A Critical Biography* (London: Oxford University Press, 1961). A definitive study, based upon exhaustive research into manuscript sources.

[143] Semple, Janet, *Bentham's Prison: A Study of the Panopticon Penitentiary* (Oxford: Clarendon Press, 1993). A fine detailed study, based on Bentham's manuscripts, which confutes simplistic readings like Foucault [56].

[144] Smith, D. W., *Helvétius: A Study in Persecution* (Oxford: Clarendon Press, 1965). A fine account of the pioneering utilitarian.

[145] Spadafora, David, *The Idea of Progress in Eighteenth Century Britain* (New Haven, CT: Yale University Press, 1990). Extremely thorough.

[146] Talmon, J. L., *The Rise of Totalitarian Democracy* (London: Secker and Warburg, 1952). Classic statement of the claim that in Rousseau's notion of freedom and the general will lay the seeds of modern totalitarianism.

[147] Todd, Janet, *The Sign of Angellica: Women, Writing and Fiction, 1660–1800* (London: Virago, 1989). A key survey of women writers in Britain, to be used in conjunction with [149].

[148] Tomaselli, Sylvana, 'The Enlightenment Debate on Women', *History Workshop Journal*, 20 (1985): 101–24. Demonstrates that many Enlightenment thinkers believed that women had played a major role in the emergence of modern, polite, advanced society.

[149] Turner, Cheryl, *Living by the Pen: Women Writers in the Eighteenth Century* (London: Routledge, 1992). Use alongside [147].

[150] Vartanian, Aram, *Diderot and Descartes: A Study of Scientific Naturalism in the Enlightenment* (Princeton, NJ: Princeton University Press, 1953). A sensitive account of the influence of Descartes in the development of materialist thought.

[151] Venturi, Franco, *Utopia and Reform in the Enlightenment* (Cambridge: Cambridge University Press, 1971). Venturi explores the attractions yet ambiguities of the republican tradition of political thinking.

[152] ——, *Italy and the Enlightenment*, (ed.) S. Woolf (London: Longman, 1972). Major essays on the Italian Enlightenment.

[153] Vereker, C. H., *Eighteenth-Century Optimism* (Liverpool: Liverpool University Press, 1967). Examines the double-sidedness of Enlightenment 'optimism', which could equally easily be treated as a form of fatalistic pessimism.

[154] Vyverberg, Henry, *Historical Pessimism in the French Enlightenment* (Cambridge, MA: Harvard University Press, 1958). Vyverberg dispels the notion that the *philosophes* were naively optimistic prophets of progress.

[155] Wade, Ira O., *The Clandestine Organization and Diffusion of Philosophic Ideas in France from 1700 to 1750* (Princeton, NJ: Princeton University Press, 1967).

[156] ———, *The Intellectual Origins of the French Enlightenment* (Princeton, NJ: Princeton University Press, 1971).
        Wade's two works are classic studies of the sources and spread of early Enlightenment radicalism.

[157] Wangermann, Ernst, *The Austrian Achievement, 1700–1800* (New York: Harcourt, Brace, Jovanovich, 1973). Important study of the interplay of government and Enlightenment ideas in the Habsburg territories.

[158] Watt, Ian, *The Rise of the Novel* (London: Chatto and Windus, 1957). Still the best account of the ideological origins of the new form of fiction pioneered by eighteenth-century writers.

[159] Weisberger, R. William, *Speculative Freemasonry and the Enlightenment: A Study of the Craft in London, Paris, Prague and Vienna* (Boulder, CO: East European Monographs, 1993). A reliable survey of freemasonry, Europe-wide.

[160] White, R. J., *The Anti-Philosophers: A Study of the Philosophes in Eighteenth-Century France* (London: Macmillan, 1970). Together with Becker's *Heavenly City* [21], the most articulate attempt to debunk the Enlightenment.

[161] Williams, Raymond, *The Long Revolution* (London: Chatto and Windus, 1961). A synoptic study of the rise of the media within bourgeois society over the last three centuries.

[162] Wills, Garry, *Inventing America: Jefferson's Declaration of Independence* (Garden City, NY: Doubleday, 1978). Wills argues that the crucial documents of the new republic owed much to the moral philosophy and political outlooks of the Scottish Enlightenment.

[163] Wilson, Adrian, *The Making of Man Midwifery* (London: University College Press, 1995). An insightful history of early modern childbirth.

[164] Wilson, Arthur, *Diderot: The Testing Years, 1713–1759* (New York: Oxford University Press, 1969). The best biography.

[165] Yolton, John, *John Locke and the Way of Ideas* (New York: Oxford University Press, 1956). A perceptive account of the revolutionary nature of Locke's epistemology.

83

Index

Calas family (Jean Calas (1698–1762) executed for murder), 35
Campanella, Tommaso (1568–1639), 23
Cassirer, Ernst, *The Philosophy of the Enlightenment* (1964), 39
Catherine II, the Great (1729–96), empress of Russia, 6, 23, 45, 47, 57
Catholicism, *see* Christianity
Charles II (1630–85), 41
Charrière, Isabelle de (*née* van Zuylen; 1740–1805), 45
Châtelet, Gabrielle Émilie, marquise de (1706–49), 45
children, feral, 16
Christianity,
    Bible, 13, 14, 15, 16, 34, 36, 61, 65, 67
    Catholic/Protestant disputes, 14–15
    Catholicism, 16, 29, 35, 48, 52, 57
    criticisms of, 29–30, 34–7
    decline of, 66–7
    in Enlightenment, 15, 16–17, 19, 20, 29–37
    history of, 66
    humanitarian movements, 62
    ideas of dying, 61
    ideas of man, 12,
    persecution, 16, 23, 30, 35, 41–2, 66
    Protestantism, 33, 35, 52
    sexual taboos, 56
Cicero, Marcus Tullius (106–43 BC), 12
Coleridge, Samuel Taylor (1772–1834), 68
Collins, Anthony (1676–1729), 33
Condillac, Etienne de (1715–80), 4, 18, 39, 68
Condorcet, M. J. A. N., marquis de (1734–94), 3, 4, 28, 40, 43, 64
    on perfectibility of man, 16, 18
Cooke, James (1728–79), 57–8

Cooper, Anthony Ashley, 3rd earl of Shaftesbury (1671–1713), 59
Copernicus, Nicolaus (1473–1543), 12
Copernican astronomy, 13; (as heresy), 16
cosmology:
    sixteenth-century, 13
    seventeenth-century, 12, 13, 16
Counter-Reformation, 2, 14, 52
criminology, 61–2
Crocker, Lester G., ix

Dante Alighieri (1262–1321), 66
Darnton, Robert, 41, 42, 44
Darwin, Charles Robert (1809–82), 18, 66
Darwin, Erasmus (1731–1802), 18–19, 21, 33, 44, 60, 61, 64
death, ideas of, 61
Defoe, Daniel (*c*.1661–1731), *Robinson Crusoe* (1719), 51, 59
Delacroix, Eugène (1798–1863), 66
Descartes, René (1596–1650), 2, 13, 14, 15
Diderot, Denis (1713–84), 3, 40, 43–4, 60, 65
    hostile to religion, 29, 36
    patronised by Catherine the Great, 6–7, 23, 47
    Rameau's Nephew (1762), 59
    Supplément au Voyage de Bougainville (1772), 56
Diderot, Denis and Alembert, Jean Le Rond d', *Encyclopédie* (1751), ix, 3, 43
discovery and exploration, 13, 16, 56
Donne, John (1572–1631), 14
Doyle, William, ix
Dutch Republic, 15, 36
    as source of Enlightenment, 41, 48–9
    politico-religious freedom in, 49
    publishing trade, importance of, 49

Helvétius, Claude Adrien
  (1715–71), 4, 11, 17, 18,
  27, 39, 43, 68
Herder, Johann Gottfried
  (1744–1803), 4, 16
Holbach, P. H. T., Baron d'
  (1723–89), 4, 8, 38, 44
  hostile to religion, 29, 30
  Système de la Nature (1770), 34
Holland, *see* Dutch Republic
Holy Roman Empire, 13
Horace (65–8 BC), 1
Horkheimer, Max, 8
Huguenots, 15, 41
humanitarianism, 2, 59–60, 62
Hume, David (1711–76), 4, 57
  attitude to religion, 31–2, 61
  moral philosophy, 11, 15,
    31–2, 59
  role in Scottish Enlightenment, 51
Hutcheson, Francis
  (1694–1746), 59

industrialisation, 20, 51
Islam, 32,
Italy
  Enlightenment in, 11, 54
  religious persecution (sixteenth/
    seventeenth centuries), 23
  Renaissance studies of man, 12

Jacob, Margaret, 41
James II (1633–1701), 41
Jaucourt, Louis, Chevalier de
  (1704–80), 42
Jefferson, Thomas (1743–1826),
  4, 62
Jesuits, 9, 29
Jesus Christ, 42
  see also Christianity
Jews, 48
Johnson, Samuel (1709–84),
  Rasselas: The Prince of Abyssinia
  (1759), 58
Joseph II (1741–90), Holy Roman
  Emperor, 52, 64
Justi, Jean Henri Cottlob (d.1771),
  27, 53

Kant, Immanuel (1724–1804),
  1, 4, 26, 39
Kaunitz, Wenzel Anton Dominik,
  count (1711–94), 53
Kepler, Johannes (1751–1630), 13

Lamarck, Jean Baptiste de
  (1744–1829), 18
La Mettrie, Julien (1709–51), 4, 11
languages,
  French, 47–8
  Latin, 47
legal reform, *see* Enlightenment
Leibniz, Gottfried Ephraim
  (1646–1716), 2, 3
Lessing, Gotthold Ephraim
  (1729–81), 60
literature
  journals, 51, 53, 68
  novels, 58–60
Livy (Titus Livius: 59 BC–AD 17),
  12
Locke, John (1632–1704), 18, 20,
  27, 50
  empiricist, 39
  *Essay Concerning Human
    Understanding* (1690),
    18, 59
  *Some Thoughts Concerning
    Education* (1693), 18
  The Reasonableness of Christianity
    (1695), 33
London Stock Exchange, 26
Louis XIV (1638–1715), king of
  France, 15, 25
Louis XV (1710–74), king of
  France, 22, 25, 54
Louis XVI (1754–93), king of
  France, 28, 54, 64
Lucretius (*c*.94–*c*.55 BC), 35
Lunar Society of Birmingham,
  43, 44
Luther, Martin (1483–1546), 12

Mably, Gabriel Bonnot de, abbé
  (1709–85), 27, 31, 65
Machiavelli, Nicolò
  (1469–1527), 13

Toland, John (1670–1722), 34, 41–2
Trent, Council of (1545–63), 12
Turgot, Anne Robert (1727–81), 3, 4, 54, 64
twentieth-century views on Enlightenment, viii–ix, 1–2

United States of America, constitution (1787), 4, 28
  Declaration of Independence (1776), 28
  see also America, North
universities, 36, 52
utilitarianism, 4, 17, 27, 65

Vico, Giambattista (1668–1744), 11
  Scienca nuova (1725), 40
Victorians, *see* nineteenth century
Volland (Louise Henriette), 'Sophie' (1716–84), 45
Voltaire (pseud. of François-Marie Arouet; 1694–1778), 3, 4, 6, 20, 42, 43, 44, 47, 48, 57, 62, 74
  appointed historiographer royal, 22

campaigns against legal injustice, 3, 35
criticises the Sorbonne, 52
hostile to Christianity, 4, 8, 29–30, 32, 34, 35–36
patronised by Frederick the Great, 7, 23, 47
political views, 24, 54, 65
praises Newton's achievement, 16
Candide (1759), 3, 6, 58
Lettres philosophiques (1733), 16, 26, 40, 51

Wade, Ira, 42
Watt, James (1736–1819), 44
Wedgwood, Josiah (1730–95), 44
Wesley, John (1703–91), 61
Winckelmann, Johann Joachim (1717–68), 60
Wollstonecraft, Mary (1759–97), *Rights of Woman* (1792), 45
women in the Enlightenment, 2n, 45–6, 61

Zuylen, Belle van, *see* Charrière, Isabelle de

Printed and bound in the United States of America